First published in September 2015

British Library Cataloguing in Publication Data
A catalogue record for this book is available from
the British Library.

ISBN 978 0 85733 770 2

Library of Congress control no. 2015939541

Published by Haynes Publishing,
Sparkford, Yeovil, Somerset BA22 7JJ, UK
Tel: 01963 440635
Int. tel: +44 1963 440635
Website: www.haynes.co.uk

Haynes North America Inc.
861 Lawrence Drive, Newbury Park,
California 91320, USA

Printed and bound in the USA

**While every effort is taken to ensure the accuracy
of the information given in this book, no liability can
be accepted by the author or publishers for any loss,
damage or injury caused by errors in, or omissions
from the information given.**

Credits

Author:	**Liz Shankland**
Project Manager:	**Louise McIntyre**
Copy editor:	**Derek Smith**
Page design:	**James Robertson**
Index:	**Dean Rockett**
Illustrations:	**Richard Parsons**
Photography:	**Pages: Liz Shankland (unless stated otherwise).**

Author acknowledgements

This is my fourth book – my third for Haynes – and it
has, undoubtedly, been the most challenging and
complex to write. In times of great stress and when
deadlines were pressing, Gerry Toms was a lifesaver,
taking so many of the everyday farming chores off my
shoulders so that I could shut myself away with my
computer. Without his help, this book would still be
half-written, so for his help and understanding, I'll
always be grateful.

My thanks also go to Kate Humble who, more than
three years ago, invited me to teach courses for
smallholders at Humble by Nature, her rural skills school
in Monmouthshire, and who wrote the foreword to this
book and its predecessor, the *Smallholding Manual*. It
was through Kate that I met Tim Stephens, a wonderful
shepherd who not only keeps his own commercial flock
at Kate's farm, but also teaches the courses in sheep
husbandry, lambing and shearing. Every time I meet Tim,
I learn more about good sheep management, and I have
definitely acquired a much better understanding of what
makes a good shepherd by spending time with him and
bombarding him with questions. If you get the chance
to attend one of his courses, you won't be disappointed.

And then there is Louise McIntyre – the project
manager for this book – who has remained supportive
and cheerful throughout the writing of the *Sheep
Manual*, despite all the missed deadlines. She probably
knows by now that I never deliver on time, and so builds
in extra buffers to save my skin. If I'm asked to write a
fourth Haynes Manual, I'll try harder – I promise!

Finally, my heartfelt thanks go out to all those
dozens of people who have provided pictures for this
book, or who have allowed their sheep to become
photographic models. My dear friend, Michele Baldock
did a terrific job helping me to source photographs for
the breeds section. It was a time-consuming and
sometimes frustrating job, but amazingly she is still
talking to me. A book like this is nothing without good
illustrations, and, thanks to the contributions of so
many helpful friends – old and new – this *Sheep
Manual* is unique when compared to other titles in the
same genre.

Haynes
Sheep
Manual

The step-by-step guide
to caring for your first flock

Liz Shankland

Foreword by Kate Humble

CONTENTS

FOREWORD

'I simply don't understand why anyone would want to keep sheep,' said my husband, with genuine bemusement. 'I know we don't know much about them, but it seems to me that they either have rotting feet, maggots in unmentionable places, do everything they can to escape or just drop dead for no apparent reason.' I remember quoting that back to Jim Beavan, the ever-patient farmer who took me on as his apprentice shepherd for the first series of the BBC's *Lambing Live*. 'He's not far wrong!' he laughed. But there must be something about them, because I now spend the autumn obsessively looking to see if the ram has left coloured splodges of raddle on the bums of my ewes – evidence that he has been at work. By April I'm exhausted and have permanent bags under my eyes because I've just emerged from the sleepless marathon that is lambing and in the summer I'm covered in lanolin and poo, with an aching back as I try to get to grips with shearing.

At the Humble by Nature farm we teach a variety of rural skills and animal husbandry courses. Liz Shankland's smallholding courses and farmer Tim Stephens' sheep-related courses continue to grow in popularity. There appears to be a burgeoning desire for people to find ways to better connect to their landscape, to utilise their land, and take some control over how they put food on the table. Keeping sheep is certainly one way to achieve all those things. Some people simply want a few weaned lambs to mow the grass and end up in the freezer. The primary interest of others is what they can do with the wool, while some plan to have a permanent flock to breed, providing animals for their own table and to sell on.

Whatever the reason you are considering keeping sheep, this book, with its honest, down-to-earth approach is an essential place to start. As a smallholder who has been through the countless ups and downs of keeping livestock herself, Liz's practical experience will give you a wealth of invaluable insight and information. There're even top tips on how to avoid maggots...

Good Luck!

Kate Humble

Liz Shankland

Josh Shankland

INTRODUCTION

Could there ever be more useful animals than sheep? As well as being the ultimate land-management tools, grazing and fertilising as they move systematically along, they bring a whole host of other benefits.

Their wool allows us to make clothes, carpets, home and industrial insulation, and numerous other everyday things. Skins can be tanned to create the softest leather, turned into sumptuous, fluffy rugs, or made into the window cleaner's most useful tool, the chamois.

Lanolin, the yellow, waxy substance produced by the sheep's sebaceous glands to keep the fleece dry, is used in everything from motor oils and printer ink to cosmetics and pharmaceutical products.

Ah, and then there is the meat. Few dishes can rival the taste of a mint-enhanced joint of lamb or a slowly cooked casserole of mouth-watering mutton. But don't forget the milk, either; a boon to those allergic to traditional dairy products, it is also the essential ingredient in many of the world's favourite cheeses, such as Roquefort, feta, ricotta, and Pecorina Romano. What's more, there is a growing market for products like butter, yogurt and ice cream.

Having said all that, sheep have their faults. They can be nervous, flighty and difficult to handle; easily panicked, they can get themselves into dreadful trouble as they recklessly run away from what scares them; and they can have a tendency to fall ill or die, sometimes without giving the shepherd the merest hint that something is wrong.

Nevertheless, the benefits that sheep bring vastly outweigh the drawbacks to keeping them. It's something on which farmers and smallholders across the world agree: sheep are worth the effort.

This book is intended as a starter manual for anyone who wants to find out more about these intriguing animals. Whether you are simply interested in keeping a few sheep as lawnmowers, establishing a small breeding flock, or maybe even setting up a business where sheep play a key part, this book will demystify the whole process, setting out what you need to do before buying your animals, how to choose the right stock and how to care for them at all the key stages of life.

Be aware, however, that you shouldn't just cherry-pick the chapters you think you need to read. Sheep are complex animals and you'll find that, for instance, if you want to know about breeding, you will need to read about feeding and nutrition first, and you will also need to read about basic health and husbandry, too, as there are many preventative factors to consider if you want to avoid problems further down the line.

Don't forget that there is no substitute for experience. You can swot up on theory as much as you like, but you'll have added peace of mind if you take time to familiarise yourself with sheep husbandry – preferably under the watchful eye of someone knowledgeable – before you have responsibility for your own flock.

Enjoy getting to know your new multi-purpose, fleecy friends, and make the most of these most valuable and versatile creatures.

Liz Shankland, 2015

Josh Shankland

SHEEP: THEIR HISTORY AND IMPORTANCE

Above: Mouflon sheep.

Below: Pub and place names show the importance of sheep in communities.

It's estimated that sheep have been cared for by man for around 10,000 years. They were among the first animals to be domesticated – followed by goats, pigs and cattle – largely because they were easy to catch and handle, were not aggressive, and because they reached breeding age early and were incredibly productive.

After all, what was the point of being a nomadic hunter-gatherer covering many miles a day in search of food when you could put down roots in one place with your family and friends? When you needed meat, our farming ancestors could just pop outside and kill an animal, which was already securely penned up, fairly tame, needed no chasing and offered little resistance.

Before settling down completely, the hunters became pastoralists – people who kept their own livestock but who moved them to better grazing whenever required. Eventually, instead of travelling endlessly, the hunters began to establish settlements, which became the first villages – and the first farms. The communities gradually learned to make the most out of the multi-purpose sheep, using its skin and wool for clothing and harvesting its milk.

The origins of sheep continue to confuse and divide historians and scientists alike, but it is generally accepted that all the breeds in existence today are descended from two distinct types of sheep – both blessed with impressive horns. The Mouflon – a hairy, rather than woolly sheep – was found in most parts of Europe, and modern-day Mouflon still bear a striking similarity to many of our primitive breeds. The Urial (*Ovis orientalis*) is thought to have originated in the countries that made up the old Ottoman Empire and is now found mostly in western central Asia. Two other types are sometimes suggested as contributing to our contemporary breeds – the Argali (*Ovis ammon*), found in central and northern Asia, and the Bighorn (*Ovis canadensis*), originally from north-east Asia and Siberia but now found in North America. Again, both have large, remarkable horns, curving outwards.

Our relationship with sheep has endured and, today the global population stands at more than one billion – which works out at a ratio of roughly one sheep to every six humans.

Domesticated sheep (*Ovis aries)* are an important part of the agricultural economy worldwide, providing meat and milk for food, wool for clothing, carpets and insulation, and numerous other useful by-products. In some developing countries, survival would be difficult without this versatile all-rounder close to hand. Sheep farming is a common practice throughout the world, largely because sheep are so easy to manage, are adaptable and require a relatively low input of feed. Think of which countries are most associated with sheep and a few come to mind. Most people would hazard a guess at New Zealand having the largest flocks, but, in fact, China holds the record, with almost 140 million sheep; Australia is pushed into second place, with a mere 98 million, followed by Iran (53 million), the former Soviet Union states (52 million) and then New Zealand (46 million). The UK, by comparison, has just 25 million.

Small-scale sheep keeping is on the increase. The most recent statistics* from Defra (Department for Environment, Food and Rural Affairs) relating to the number of sheep

Michele Baldock

Above: Sheep can be companion animals too.

Below: These specially-trained sheep entertain crowds at agricultural shows.

Above: Urial rams go head-to-head.

farms/smallholding in the UK make interesting reading: at the time of the survey, there were almost 10,000 smallholders with small flocks of sheep. It is worth noting that, between 2005 and 2011, whereas the number of commercial sheep holdings fell by 12.8%, the number of smallholdings with fewer than 20 sheep rose by 4.2%. The figures suggest that, regardless of market trends, there remains a sustained growth in the practice of keeping sheep in small numbers in the UK.

Wool prices are improving, too. For years, small-scale shepherds who had to pay for a shearer to come in actually lost money on their fleeces, such was the poor price paid, but increased demand for wool for a whole range of products has made it possible, once again, to turn a small profit. All of this, of course, is a far cry from the days when wool was one of the most valuable forms of currency. Back in the Middle Ages, for instance, few young and healthy sheep were slaughtered for meat because their fleeces were worth many times more than the entire animal would make if sold for food.

Back in the 12th century, the Cistercian monks – intrepid pioneers of self-sufficiency who travelled vast distances building abbeys and establishing farmsteads to keep the brotherhood fed and clothed – became some of the most successful shepherds of their era. It's been

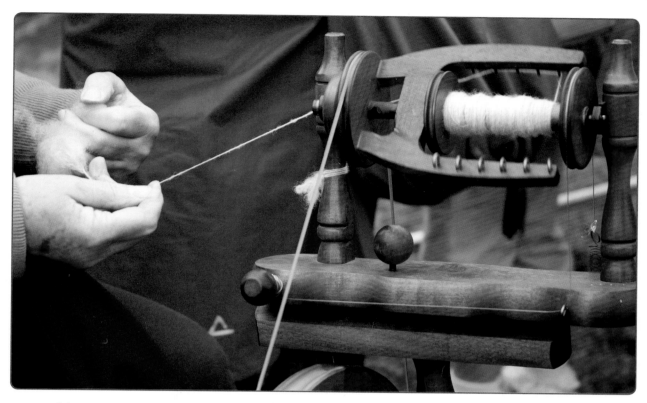

suggested that, across Britain, they reared flocks amounting to a million sheep. They kept them partly to make vellum – parchment used for writing on before paper came along – and partly for wool, which they used for clothing and also exported for good money. A papal agreement meant they didn't pay any taxes, no matter where in the world they settled, so they did far better than any other wool producers of their time.

With such a lucrative product at their disposal, it's no wonder the Cistercians chose to harvest wool from their sheep rather than meat. They undoubtedly ate well at their settlements, but lamb or mutton would never have been a regular dish on the menu – it would have been far too wasteful to have killed a sheep aged a year or two old when it had the potential to go on producing a regular supply of valuable wool for many years to come. Wool emerged as a prized export for entrepreneurial Europeans and, as a thriving textile industry developed quickly over the centuries, so numerous towns and cities were established and made prosperous as a result.

Today, meat is the prime reason for rearing sheep. Exploding human populations have pushed up the demand for lamb and mutton worldwide. With faster-maturing,

Above and below: Our ancestors valued sheep more for their wool than their meat.

highly fertile breeds available that are capable of giving birth to multiple lambs instead of singles, flock replacements can now be raised quickly and cheaply. So now, you can have your sheep and eat it, too.

*Agriculture in the United Kingdom 2011 (www.defra.gov.uk)

GETTING STARTED

Do you really want to be a shepherd?

Before you take the plunge, it's worth asking yourself a few questions.

Why do I want to keep sheep?

Do you simply like the *idea* of sheep? Are you really keen on raising your own meat, spinning your own wool and getting into the breeding side of things, or do you just need something to graze your paddocks, keeping the grass and scrub at bay? You may simply want to enhance the view from your kitchen window. After all, estate agents often say that the sight of a small flock of sheep grazing in a paddock in front of a rural property can add £10,000 to the value!

Above: Students on a lambing course.

Don't forget that there are other ways of having the benefit of sheep without actually keeping a conventional breeding flock. You don't have to throw yourself completely into shepherding in order to enjoy having sheep around. Some options to consider:

■ Allow a neighbouring farmer to graze your land with *their* sheep – either for payment or maybe in exchange for meat or other services, such as help with your land. This way you get to see sheep every day, you can learn about their behaviour and their needs, you can offer to lend a hand when the owner has to carry out routine tasks like foot trimming, shearing and perhaps even lambing. While shepherding from a distance, you may decide that sheep keeping is not for you – in which case, you might have saved a lot of time, effort and money. You can either have sheep grazing all year round or just take in 'tack' sheep (see Glossary) over winter.
■ Buying in sheep to raise for meat gives you dual benefits – the pleasure of keeping sheep for a few months of the year, without a long-term commitment, plus home-grown meat at the end of the process. You can buy 'store' lambs – weaned lambs, which will need a few months more growing time before slaughter, and you may also find orphan or 'pet' lambs available around which need bottle-feeding. See more on this in Chapter 11.
■ Keeping a non-breeding flock means you will never have the stress of lambing time. If you just need a few friendly lawnmowers, get yourself a few wethers – castrated rams. Opting for males means you won't get tempted into breeding when you might not be ready for it. If you buy females, you may be tempted to start breeding when you're not really ready for it – and there's always the chance that a neighbour's ram might jump the fence and get them in lamb when you least expect it.

How much do you know about sheep husbandry?

Doing your homework is essential before you take on any livestock. Although no one expects you to have an in-depth knowledge of sheep health problems, you should be aware of the kind of regular care and attention they are likely to need – which is why getting some experience with someone else's animals can be so valuable. It's so much easier doing something for the first time under the supervision of someone with experience than flying solo, wondering whether or not you're doing it right. If you can't find anyone locally who is willing to let you help out, there are plenty of courses available.

Are you physically fit for the challenge?

Livestock not only need to be fed and watered, they also require regular checks on their general health and well-being. Depending on where you keep your sheep, this can be hard work, particularly through the winter months. Can you see yourself braving the wind, rain and snow every day, maybe carrying sacks of feed or bales of hay? Would you be able to catch and turn over (see Chapter 5) a lame sheep to examine its feet? What if one fell down a ditch and had to be hauled out? Could you manage that? Also, what would happen if you fell ill or had an accident? Do you have someone who could step into your shoes?

How emotionally prepared are you?

Have you heard the old farmers' saying, 'Where there's livestock, there's dead stock'? Looking after animals has high and low points and there will be times when you will be faced with illness, injury and death. When you're just starting out, you won't want to think of worst-case scenarios, but you need to consider how you might cope, for instance, if a favourite sheep had to be euthanised. Lambing time, too, can be marred by dramas and disappointments, and while most births will be straightforward, you have to prepare for stillbirths, deformities, and sometimes deaths of both ewes and lambs. Finally, if you intend raising sheep for meat, are you convinced you could take them to slaughter when the time comes? If you're not sure, it might be best to rethink your plans now.

How secure is your land?

Make sure that all boundary fences and hedges are sheep-proof and dog-proof. Sheep will soon find a way through wobbly fencing or a gappy hedge that hasn't been properly maintained. The last thing you want is for your flock to wander into a neighbour's garden or, worse still, on to a public road. Similarly, you will want to keep other people's dogs out in order to keep your sheep safe. Finally, it is worth considering taking out a specialist farm or smallholding insurance policy, in case of third-party accidents.

Do you have sufficient space?

Recommendations on stocking densities vary. Although the National Sheep Association says six to ten adults, plus their lambs, per acre, this recommendation applies to good, well-kept, highly productive grassland, which can provide plenty of nutritious food. Maintaining grassland to a high standard takes hard work and experience and it is more likely that the average small farm or smallholding will have less-than-perfect grazing. For this reason, the most common recommended stocking density is four or five adults plus their lambs to each acre.

Climate, topography and soil type will all have an impact on how well your grass fares, and therefore how much food will be available. If in doubt, understock initially and see how things go. Talk to neighbouring farmers with similar land about their grazing methods and the numbers of sheep they would recommend. Local knowledge can be invaluable when you are starting off.

There may be times of the year when you need to take your sheep under cover – for instance at lambing time or if the ground becomes too waterlogged. You may also need to treat sick sheep and give them bed rest. Many flocks spend all their time facing the elements, but a barn, shed or polytunnel can make life much easier – for sheep and shepherd alike.

Below: Escaped sheep can wreak havoc on a vegetable plot.

Legal requirements

Whether you intend to have just two sheep or 2,000, the rules are the same. There are strict rules and regulations surrounding the raising of livestock and the movement of animals between farms, smallholdings, markets and abattoirs, and your first move must be to register your land as an agricultural holding.

Above: Tags and tagging pliers.

Step 1: Getting registered

The process of registering your holding is similar in England, Scotland and Wales but varies slightly in Ireland (see Useful Contacts).

In England, Scotland and Wales, you need a County Parish Holding (CPH) number, which identifies your land as a holding and allows you to legally keep livestock. The number will look something like this: 58/421/0086. The first two digits are the county, the next three the parish in which you live, and the final four the actual number of your holding. You will use this unique number whenever you buy or sell livestock, move animals on or off your premises, when ordering identification tags, and in various official documents.

Ireland doesn't use CPH numbers, but anyone wanting to keep livestock will still need a flock number (see below).

Getting a CPH isn't difficult, doesn't cost anything and shouldn't take too long. In England, you'll need to contact the Rural Payments Agency; in Wales, it's the appropriate divisional office of the Welsh Assembly Government; and in Scotland it's the Rural Payments and Inspections Department. Just make a phone call and you'll be sent some very straightforward paperwork to fill in and return.

Step 2: Get a flock number

Once you have the CPH, contact the Animal and Plant Health Agency (APHA) for a flock number. You can do this after you have received your first sheep, but it's easier to do it

Below: Sheep which have been double tagged.

beforehand. Your flock number will be used on all documentation whenever you move the animals to another location or to slaughter and needs to be imprinted on ear tags.

In Northern Ireland, flock numbers are obtained from the Department of Agriculture and Rural Development, and in the Republic of Ireland, it's the Department of Agriculture, Food and the Marine.

Step 3: Identification, tagging and movement

As soon as you have your flock number, you will be able to order identification tags for your animals. Tags get cheaper the more you order, but most companies will offer to sell in small batches.

The tagging system for sheep has been notoriously complex and changeable in recent years. To make matters more confusing, there have also been differences in the way in which the devolved administrations in the UK have chosen to deal with things. At the time of writing (2015), the following rules apply, but do check with the relevant authority in case of post-publication changes.

Tagging must take place before sheep reach nine months of age (before six months if housed overnight), or before they leave the holding of birth – whichever comes sooner.

- **Sheep over 12 months old.** Must have two identifying tags bearing the flock number; the one in the left ear must be a yellow electronic identification tag (EID), which contains a microchip holding the animal's unique ID number. The information can be read and transferred to a computer using a handheld scanner (e.g. at markets or abattoirs). The non-EID tag (often referred to as the 'visual' tag) can be any colour, apart from yellow.
- **Sheep under 12 months of age being reared for meat.** These must be identified with a single EID tag.
- **Sheep tagged before the EID system started.** If you have older animals on your holding, which were already tagged before 31 December 2009, there is no need to electronically identify them. The only exception is if they subsequently move from your holding. Then they should be double-tagged (EID).

Above: Traditional movement forms.

- **Pedigree sheep.** In addition to the standard tagging requirements, breed societies will have their own rules on identification in respect of pedigree breeding stock, so check with them regarding tattooing or notching.
- **Lost tags.** Tags have a habit of getting caught on fencing and hedges. If a sheep loses one of its tags, you must either order a replacement from the tagging company that supplied it or remove the remaining tag and insert a fresh pair – remembering to make a note of the change of number in your flock register (see page 22).

Moving sheep

Keepers must inspect sheep for signs of disease before movement. Any signs of the more serious notifiable diseases must be reported (to the regional branch of the Animal and Plant Health Agency in England, Scotland and Wales, or to your local veterinary office of Department of Agriculture in Northern Ireland and the Republic of Ireland).

All sheep, goat and deer movements – whether to another farm, to an abattoir, a market, or a show – must be accompanied by a completed animal movement licence (AML1). Traditionally, this was a paper document, but there are now online reporting systems, too. Wales is due, in 2016, to join England, Scotland and Northern Ireland in having an electronic system, leaving only the Republic of Ireland still favouring the paper-only option.

Some leeway has been allowed for those who don't have access to computers, or who wish to continue using paper licences until their stock of forms runs out. Details must be completed by both the departure premises (i.e. where the animals were kept) and the destination premises. The person sending the animals must confirm that a movement has taken place, and the person receiving must do the same when they arrive.

Right: Online reporting system for England.

HOW TO REPORT A MOVEMENT

Registering for online movements is relatively straightforward – you will need to input all your contact details, CPH number and flock number – and all services have telephone helplines in case of problems. If you are buying stock, the seller will organise the start of the movement process and you will then be required to confirm the completion of the movement when you have received the animals. Confirmation must be sent within three days of receiving stock, whether online or paper systems are used.

How it works in different areas

- England: Animal Reporting and Movement Service (ARAMS). Paper forms can still be used.
- Northern Ireland: Animal and Public Health Information System (APHIS)
- Republic of Ireland: No online system as yet.
- Scotland: ScotEID. Paper forms can still be used, too.
- Wales: EIDCymru. Online service due to start in 2016. Paper forms still in use.
  See page 182 for contact details.

Standstill

Reducing the risk of disease being spread is paramount. For this reason, any movement on to your land triggers a 'standstill' – a quarantine period that varies from species to species. When you move any new livestock on to your holding, animals that were already there aren't allowed to leave until the relevant time period has expired. The only exception is for animals that are being sent direct to slaughter. Sheep, goats, cattle and deer aren't allowed to move for six days, while pigs are restricted for 20 days. The only way to avoid standstill is to either carefully plan your movements or to have an approved isolation unit (see Chapter 6).

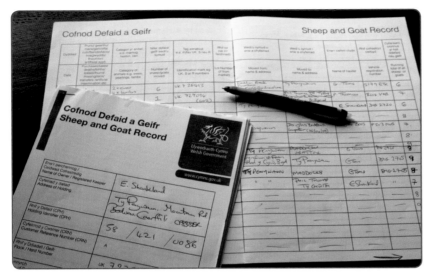

Above: Flock register.

Keeping a flock register

All sheep and goat keepers are obliged to keep a flock register, which contains details of animals kept on the holding, movements on and off, births and deaths. These details can be kept in a variety of forms. You can create a computer-based register or keep notes in a book, for instance. Depending on where you live, you may be given a register when you apply for your flock number. Local Authority Trading Standards departments will also supply sheets for you to fill in and store in a file.

In addition to keeping your own records, you will also be sent an annual sheep and goat inventory to complete, in which you will be required to list details of how many animals you currently have on site.

Keeping veterinary records

If you keep livestock, you must keep a record of any veterinary medicines administered. Records must be kept for five years and be made available for inspection on request.

Trading Standards departments will supply forms for you to fill in and place in a file, or you can store records on computer or in a notebook. The kind of details required include the date the medicine was purchased, its name, batch ID and quantity purchased; the name and address of the supplier; the dates treatment started and finished; the date the withdrawal period ended; ID numbers of animals treated; the total quantity used; and the name of the person who administered the medicine. If a product is disposed of before being finished (e.g. if it goes out of date), details of the date and

place disposed of and the quantity involved must be recorded.
For more information, see the Veterinary Medicines Directorate website, www.vmd.defra.gov.uk

Transporting sheep

In the European Union, there are extensive rules regarding the welfare of animals during transport. Most apply to everyone – whether commercial farmers or 'hobby' sheep keepers – but there are some additional regulations that are specifically aimed at people who are moving animals in connection with an economic activity.

THE BASICS

- Animals must not be transported in a way which is likely to cause them injury or undue suffering.
- Journey times must be kept to a minimum.
- The animals must be fit to travel.
- Those handling animals must be competent to do so.
- The vehicle and its loading and unloading facilities must be designed, constructed and maintained to avoid injury and suffering, and to ensure safety.
- There must be sufficient floor space and headroom.
- Animals of different ages, sizes and sexes are grouped accordingly.
- Water, feed and opportunity to rest must be made available as appropriate.
- Documentation must be carried which states the holding of departure; ownership; destination; date and time of departure; and expected duration of the journey.

TRANSPORTING IN CONNECTION WITH AN ECONOMIC ACTIVITY

Any movement of livestock that is part of a business or commercial activity, or which results in financial gain, is classed as an economic activity.

An EU regulation that has been in force in January 2007 means that anyone transporting live vertebrate animals more than 65km (40 miles) in connection with an economic or commercial activity must apply for a Transporter Authorisation. This can be obtained free of charge from the Animal and Plant Health Agency and must be renewed every five years.

The regulation does not apply to transporting animals to or from veterinary practices on veterinary advice.

There are two kinds of Authorisation – Type 1 is for short journeys (more than 65km and up to eight hours long), and Type 2 for longer journeys (more than eight hours).

For both types of journey, details of each journey must be recorded. You can either fill out an Animal Transport Certificate, available from the APHA, or record it in a book or

computer file. The information does not have to be sent anywhere, but must be kept on file for six months.

CERTIFICATE OF COMPETENCE

In addition to the Transporter Authorisation, drivers and attendants accompanying vertebrates must have a Certificate of Competence. To obtain a certificate (and again, there are two types – one for short and one for longer journeys), you need to undergo an assessment at an approved examination centre – usually an agricultural college or another college of further or higher education. The assessment is a series of multiple choice questions aimed at testing your knowledge of animal welfare, particularly with regard to health and safety during transportation. It's a City & Guilds qualification and there is a fee payable, but livestock keepers who satisfy certain farming criteria may be able to apply for funding. A separate test must be taken for each species you wish to transport.

TRAILER REGULATIONS

Animals have to be loaded, transported and offloaded in a safe and hygienic way. The obvious way is in a purpose-built livestock trailer or lorry, but many four-wheel-drive vehicles can be fitted with purpose-built livestock canopies, which meet the various regulations.

RULES RELATING TO RAMPS

There are specific EU regulations relating to ramps. For sheep and adult cattle, the angle should be no steeper than 50%. The best method of establishing the slope of your trailer ramp is to park it on flat ground with the trailer ramp down as shown in the illustration.

Measure the height of the trailer ramp as shown at point A, the ground length, marked B, and the tailgate length, C. The percentage slope is calculated by the height divided by the ground length and multiplied by the tailgate length.

In the example shown, the trailer height is 30cm, the ground length is 100cm and the tailgate length is 110cm – giving a slope of 33%. The calculation result is below the maximum of 50% and is therefore legal for the transportation of all animals. Modern trailers are designed to comply with these regulations. If you have an older-style trailer, it would be worth carrying out the calculation test to see whether it complies with the current rules.

Trailer angle calculation $\dfrac{A}{B} \times C$

A = Height
B = Ground length
C = Tailgate length

Example
A – Height $= 30$cm
B – Ground length $= 100$cm
C – Tailgate length $= 110$cm

Calculation $\dfrac{30}{100} \times 110 = 33\%$

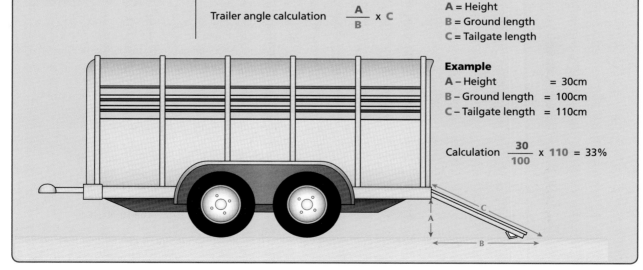

PREPARING THE LAND

Whatever kind of sheep you decide to buy, the basic requirements for keeping them will be the same. As mentioned earlier, making sure your boundaries are secure is vitally important – to keep your sheep in and unwanted visitors (two- and four-legged) out.

Stock fencing is the most popular way of providing a safe environment for your flock, but it needs to be done properly. It's a labour-intensive job and can be an expensive outlay, but there's no real alternative to it if you need to make your fields secure in a short space of time. For a detailed, step-by-step guide to fencing, see the Haynes *Smallholding Manual*.

Electric fencing – used properly – can be a quick fix and a cost-effective option, whether used as main perimeter fencing, as a backup to stock fencing, or for strip-grazing a large area. Sheep will need several strands of wire, three at the minimum, but five is better. As the sheep's wool is an

Below: Electric fencing is quick to erect and effective

Below: At least three strands of electrified wire are needed for sheep.

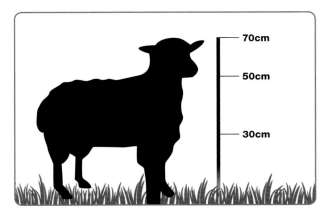

70cm

50cm

30cm

insulating substance, multiple lines are essential to make sure they feel the shock of the fence.

Suppliers of electric fencing kits will be happy to advise you on specific needs, but most have a lot of general advice on their websites. Don't rule out electric fencing because you don't have mains power in your fields, as lots of systems will operate efficiently powered by a car battery and/or solar energy.

If you decide to make a temporary lambing enclosure using electric fencing, make sure the bottom strand is sufficiently low to prevent new additions escaping.

A combination of standard stock fencing plus electric wire is another popular option. Insulators are screwed into the existing wooden posts and electrified wire is fed through them.

An alternative to wire is electrified netting similar to the sort used for poultry. Fairly light to handle and easy to install, it also offers better protection from predators. Most kits come with built-in posts, which makes erection and moving on to new ground a lot easier. In addition, the square pattern creates more of a visual barrier – to sheep and predators alike. As unshorn sheep can be well insulated against electric shocks, a powerful energiser is essential.

Living barriers

Well-laid hedging not only provides a way of containing stock but also gives shelter for livestock from harsh winter weather to blisteringly hot summer sunshine. Hedgerows provide a source of food and a habitat for wildlife, as well as acting as 'green corridors', allowing animals to travel in safety between suitable areas of habitat.

Hedgerows require regular maintenance; without being

Above: If well-maintained, traditional hedges are good barriers.

maintained, branches will simply grow upwards and develop into trees, leaving gaps below through which livestock can escape. Maintaining hedges can be time-consuming, but immensely satisfying, and it should be easy to find a course to teach you the basics. The same goes for drystone walling. Local authorities, wildlife conservation groups and farming unions often organise their own courses, or promote ones run by other organisations.

If you don't feel up to such jobs, ask around the area, as enterprising farmers often turn their hand to contract hedging, walling and fencing during quieter times.

Below: Gappy hedges provide easy escape routes.

Good grassland – a valuable commodity

For most livestock owners, grass is the most important and most cost-effective crop of all. It's a common misconception to think of it as a 'free' food, because it needs time and effort to ensure that it stays healthy and productive.

There are two things over which you have no control: the location of your land and the weather. A hilltop holding in an exposed position in an area prone to high levels of rain will never have grass to rival lowland farms on good quality soil.

Generally, agricultural land used for grazing livestock is divided into two types, permanent pasture and rough grazing. Permanent pasture is also known as 'improved' land, because it will have been cared for on a regular basis, typically having been fertilised, ploughed, drained and maybe re-seeded. Permanent pasture accounts for around three-quarters of the agricultural land in the UK. Rough grazing is unimproved land, which is largely left to its

Above: Good pasture management can save you money on feed bills.

own devices, and consequently offers less nutritional value to livestock but is often better for wildlife. Primitive breeds and hill or mountain breeds can survive on rough grazing, which includes coarse grasses and scrub, but the more modern and imported breeds can be more demanding. Even those that can manage on rough grazing will need additional feed in the run-up to lambing and during lactation.

Start with the soil

Improving soil structure is the starting point to making the best of what you have. Good soil structure helps increase its water-holding capacity. Root growth and development will be better, resulting in a healthier, faster-growing crop. Physical damage to the soil – compaction or smearing due to constant trampling by livestock or use of heavy machinery when the ground is waterlogged – has a detrimental effect on aeration and drainage and can lead to restricted root growth and a reduction in uptake of nutrients. The result can be poorer grass with a shorter growing season, loss of nutrients and soil erosion. Perhaps most importantly, it can reduce the amount of time your sheep can be kept in one place.

Below: Neglected grassland.

Testing soil status

If you are aiming to improve the overall health of your soil, there are a few tests that will help determine its status so

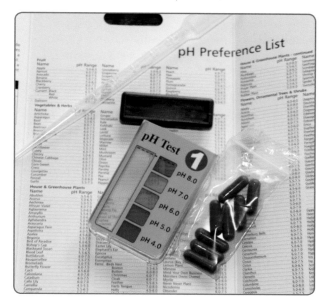

that you can take action to correct any deficiencies. The first is a pH test. The pH scale runs from 1 to 14 – with 7 being 'neutral'. Anything lower than 7 is acidic – and the lower the numbers go, the more acidic the soil. Numbers above 7 are considered

alkaline – with alkalinity increasing as the numbers rise. The optimum pH level for grass growth will vary according to soil type, but generally, around 6 is the target (the range can be from 5.6 to 6.8).

Basic home-testing kits are available in agricultural supplies stores and garden centres, but you may want to get a professional to carry out a full soil analysis and supply recommendations of what to do next.

Soils can be made more alkaline by adding lime. If your soil needs liming, it may be worth asking neighbouring farmers to do your fields when they do theirs, or hiring a contractor to do the job. Lowering the pH level of the soil to make it more acidic can be achieved, using sulphur or acidic fertilisers and manures, but it will take some time before you see results.

Soil deficiencies

Tests can be carried out to determine whether soils are deficient in nutrients or minerals. Deficiencies in nitrogen, phosphorus and potassium (abbreviated to their respective chemical symbols, N, P and K) are the most common problems, but are easily corrected. Agricultural suppliers sell a range of specialist fertilisers to treat each deficiency, and there are organic products available.

Soil should be tested every seven to eight years in fields that are kept for grazing but every four to five years if fields are cut for silage (see page 33). In areas of high rainfall, or where soils are sandy and very free-draining, testing should be more frequent – every two to three years.

Restoring neglected grassland

The productivity of pasture land can be improved by introducing specialist grasses and/or white clover into the existing sward, using over-seeding or direct drilling techniques. This is cheaper and faster than reseeding the entire area – but first you have to make sure that the poor growth is not due to soil deficiencies, compaction, or overgrazing.

Weed control (see page 31) is essential prior to reseeding. Attempting to control weeds in a sward that has not been

WHAT'S A 'SWARD'?

When reading up on grassland management, you will often find references to caring for the 'sward'. All this means is 'pasture' – an area covered with grass and/or other plants, which is used for grazing livestock. The sward can contain dozens of species, including wild flowers and grasses and either native or cultivated clover. On land that has not been regularly cultivated, annual ryegrass is likely to be a feature, too. Low-growing and often found growing in areas where feet have poached the ground (e.g. gateways), it is pretty unappealing to livestock and offers poor nutrition.

properly managed can be a tough and soul-destroying job. Sometimes, complete reseeding is the only viable option, but the cost of ploughing, buying seed and sowing can be high.

With grazing stock in mind, the aim is to produce a sward that is nutritious and long-lived. While maintaining a sward that is capable of producing a good yield all year round may not be possible, careful selection of certain grass and clover seeds can help achieve better results than just opting for a single grass variety. Grasses grow particularly well in late spring and early summer, while clovers are at their best in mid-summer. Although grasses tend to have higher annual yields, they offer less in terms of protein than clovers, and therefore a combination of the two provides a good balance. In addition, clover also takes nitrogen from the atmosphere and fixes it into the soil – not just for its own use but also for the benefit of grasses growing around it. This, in turn, can reduce the need for nitrogen-based fertilisers.

Agricultural supplies stores and specialist seed producers will be happy to advise you on choosing the right seed mix for your land. A phrase you will often see on seed sacks is 'heading date'. This is the date the plant matures, which is normally about four weeks before flowering. A good pasture mix is likely to include some of the following:

- **Italian ryegrass** – Early to grow, reaching maturity mid to late May. High-yielding and quick to establish. Has a long growing season and contains high levels of sugar. Requires tight grazing to keep in check and maintain quality.
- **Hybrid ryegrass** – A cross between the Italian and perennial ryegrass, it shares characteristics of both parents. Similar in yield to the Italian ryegrass, it is, however, longer-lasting and is better for ground cover – and so better for grazing.
- **Perennial ryegrass** – The basis (along with white clover) of much of the permanent grassland in the UK, it grows easily in a variety of situations. Lower yielding than Italian ryegrass and hybrids, it is, however, longer-lived and also more flexible in a wide range of situations because it can be grazed or cut for hay, haylage and silage.
- **Timothy** – Slow to establish and late-maturing, it is, however, hardy, and better-suited to wetter, heavy soils

and winter sheep grazing pastures. It is less nutritious than the ryegrasses, having a lower sugar content.
- **Cocksfoot** – Early-maturing and productive but coarse and not very palatable. Drought-resistant, it is useful on dry, light soils.
- **Red fescue** – Useful on hill farms with poor growing conditions, it is an early-growing, hardy grass, which needs tight grazing to maintain leafiness and quality.
- **Meadow fescue** – Less vigorous and lower-yielding than ryegrasses but leafy and nutritious. Often grown with Timothy and clover. Suitable for less-intensive farms.

Clovers

As well as fixing nitrogen, these legumes are exceptionally nutritious and high in protein (27% in the case of white clover, compared to just 17% in perennial ryegrass) and therefore often make up 30 to 50% of the sward. Clovers can be used for grazing or as part of a silage crop. Red clover is suitable for short-term grazing or cutting. It has just a single growing point, so is not suitable for close or winter grazing. It is also relatively short-lived, averaging three to four years. This plant has a detrimental effect on oestrogen, and, therefore, fertility, so ewes should not be grazed on it for six weeks before, during and after tupping.

White clover is the most important forage legume grown in the UK. There are numerous varieties available, with varying leaf sizes. Smaller-leaved varieties are more tolerant of close

MAKING A DIFFERENCE

Successful grassland management can be a complex and time-consuming activity. If you don't feel up to the task of managing your land yourself, do consider seeking professional advice and hiring experienced contractors to help get the land in shape. They may suggest:

- ploughing to create better drainage and break through compacted layers;
- aerating the topsoil, using a spiked roller to break the surface, allowing the soil to 'breathe' and water to penetrate. This helps root growth and nutrient uptake;
- adding lime or organic matter to improve soil stability and encourage earthworm activity;
- installing and maintaining drains, because waterlogged soils are more prone to damage; and
- applying appropriate fertilisers to restore deficiencies in the soil.

Paying for professional help may seem an unnecessary expense when you are just starting off, but if you are serious about getting the best nutritional value out of your grass, it could be something worth considering.

grazing; medium-leaved varieties are more productive, particularly when cut for silage or hay; large-leaved strains are only used for cutting because, although they give slightly higher yields, they are less persistent when grazed.

Controlling common weeds

Weeds reduce the quality of the sward and compete with the more nutritious plants, often crowding them out. Nettles and thistles discourage grazing and can make hay and silage unpalatable. More than simply being a nuisance, some weeds are also poisonous and must be eradicated for the safety of your livestock. In addition, several weeds are listed in the Injurious Weeds Act, which requires landowners to prevent them spreading, e.g. common ragwort, spear thistle, creeping or field thistle, broadleaved and curled dock.

Poor management is one of the reasons why weeds take control. This could mean low nutrients, acidic soil, over- or under-grazing, soil compaction, or poorly seeded or poorly composed swards, resulting in lots of open patches.

To determine whether you have a weed problem, measure out a few 50cm x 50cm squares in various places in a field. If you find that more than 10% of a square has weeds, then productivity is being adversely affected, and you need to take action. Herbicides are available, but other control measures may also be used.

BUTTERCUP

Classic indicators of low soil fertility, particularly nitrogen (N), buttercups thrive in wet conditions. Too much buttercup in a hay field will reduce the quality of the crop. Improving drainage will help control it, as will increasing grass and/or clover in the sward, but herbicides may have to be used in severe cases.

Dr Lizzie Wilberforce

CHICKWEED

Probably the most widespread pasture weed. Seeds throughout the year, filling any bare patches, and is very

tolerant of cold weather, so a long-lived plant. Presence in silage affects the fermentation process, reducing silage quality. Large quantities may cause digestive upsets in grazing lambs (and calves). Control by heavy grazing plus harrowing in the autumn, and by introducing clover to fill gaps.

THISTLES

Topping (see Glossary) helps deter thistles, but, as they spread by seeds as well as roots, they are difficult to control. Frequent topping and herbicides are the recommended options.

DOCK

Even after cutting, seeds can remain dormant and viable for many years. Grazing, topping and herbicides are recommended. Herbicides should be applied to the 'rosette' stage of the emerging plant, in spring or autumn, following topping and grazing.

RAGWORT

Highly toxic, particularly when found in silage (see page 33) as livestock can't avoid eating it. Although not very palatable when growing, it may be eaten if grazing is poor. It becomes more appealing when dead or dying, due to the release of sugars. Toxins cause cirrhosis of the liver and there is no known cure. Landowners are required by law to control this plant. The

Dr Lizzie Wilberforce

best course of action is hand pulling – taking care to wear gloves and a mask. Ragwort sets seed even after uprooting, so dispose of plants safely.

NETTLES

Spread by seeds and roots, nettles are difficult to control and thrive in rich soil. Topping helps but will not stop the spread, and they must be controlled when actively growing. Spraying with suitable herbicides is likely to be the best course of action, as the long, thread-like roots break easily and regenerate.

Your year-round guide to grassland management

This is a basic guide to care and maintenance, but geographical differences mean you will need to adjust certain things to suit your particular situation. Your location, climate and other individual circumstances may alter the time of year you do things, or the way you do things.

Time of year	Action	Why?
LATE WINTER	Carry out a soil test to establish pH and other nutrient levels.	
	Optimum soil pH for grass growth is 6–6.5. Soil that is too acidic will need an application of lime to correct the pH. Potash, phosphate and nitrogen are also needed for good growth. Deficiencies may mean a compound fertiliser is needed.	
	Clear ditches and check land drains for blockages.	Taking action now saves remedial work later on.
	Plough if you plan to reseed in the spring.	To create drainage channels and ease compaction.
EARLY SPRING	Harrow.	Harrowing tugs out all the dead grass/thatch so that the soil can receive air, water and nutrients more efficiently. This should be carried out in the early spring, before the grass starts to grow in earnest.
	Re-seed if necessary (or maybe delay to late spring, depending on climate).	To cover bare patches or poorly covered areas.
	Roll if necessary.	This will repair damage caused by trampling. Roots can more easily make contact with available nutrients.
	Apply nitrogen, phosphates or potash fertilisers where necessary (again, may be better left to late spring in some areas).	Results of the soil test will have indicated whether fertilisers are needed. Timing of the application is crucial; conditions need to be right for nutrient uptake, and to prevent nutrients being washed out of the soil by rain.
LATE SPRING/ EARLY SUMMER	If fields are not being grazed sufficiently, cut with a tractor and topper or a tractor mower and maintain a length of about 5–7cm (2–3in). If the land is grazed sufficiently in April, May and June there may be no need to cut.	Grass at this length can make its own nutrients more efficiently, and is also more resilient against the impact of hooves.
	Target invasive weeds like dock, nettles, Japanese knotweed and ragwort.	Removal earlier in the year is easier than waiting until they have become stronger and more established.
SUMMER	Make hay, haylage, or silage.	Grass will soon be in flower. Cut now before it becomes 'stemmy' and fibrous and, therefore, less palatable and less digestible. Cutting and storing now for winter feed means nutrients are retained. Grass has less nutritional value once it has stopped growing.
	If not cutting for hay/silage/haylage, continue mowing or grazing with livestock to keep grass down to 5cm (2in) in length.	Helps stop weeds from flowering and setting seed.
MID/LATE SUMMER	Take a second cut for hay/silage/haylage if the grass remains sufficiently good.	Provides more over-winter feed.
	Re-seed bare patches if necessary.	Prevents weeds taking hold; sometimes conditions are better than if re-seeding earlier in the year.
AUTUMN	Continue mowing or grazing.	Helps to keep vigorous autumn growth in check.
	Carry out field maintenance tasks, e.g. repairing fences, laying hedges.	It's worth completing jobs while the weather is good.
WINTER	Keep livestock off wet fields and rotate to other fields as necessary.	Minimises the risk of poaching and compaction by livestock and allows grass and ground to recover. Sheep are best kept off wet fields to minimise the risk of fluke, while rotating also reduces risk of foot rot.

Above: Hay being loaded onto a lorry.

Hay, silage and haylage

Keeping livestock in good condition during winter means providing alternative sources of food, which can be costly. For this reason, generations of farmers have chosen to harvest grass and cereal crops during the summer months, when they are at the most nutritious stage in their development. The crops are stored and preserved by making hay, silage or haylage. Care should be taken in feeding any preserved fodder, as mould and bacteria can be extremely dangerous to ruminants. Listeriosis (see Chapter 4 and Chapter 7) is a particular risk in poor-quality or badly stored silage. Any baled fodder should be used quickly after being opened, as it deteriorates rapidly.

Hay

Grasses and other fodder plants are cut just as they are coming into flower, and then dried before being baled and stored as over-winter feed. Drying is essential for the nutritional quality of the hay but also for safety reasons. Freshly cut forage is not dead; plant cells continue to burn sugars to produce energy (the process of respiration) and heat is released in the bale – sometimes so much that a fire can start.

Haymaking can be a precarious business, as it is so weather-dependent. The traditional way is to cut the grass and leave it to dry thoroughly before baling. A hay crop can be ruined by moisture, as mould and toxins can develop, which are potentially lethal to livestock.

Silage

In many parts of the UK, silage making is more popular than hay. Grass (or other crops, such as maize) is cut and left for just a day to wilt – but not completely dry out – and then either baled in airtight wrappers or stored in a silage clamp. The harvested crop then begins to ferment. If packed into a silage clamp, the crop is compressed by driving a tractor back and forth over it to remove as much air as possible. Then it is covered with plastic silage wrap and weighed down, often with tyres. The less air in the silage, the more efficient the fermentation process, resulting in a higher-quality fodder.

Haylage

Not as moist as silage, and not as dry as hay. The crop is wilted rather than dried out completely and then baled and wrapped or bagged, as with silage. If made properly, it can be more nutritious than hay.

Using forage root crops

Forage root crops come into their own after the main cereal or grass crops have been harvested. They are often used by farmers as 'catch crops' or 'cover crops' – i.e. crops sown to grow straight after one crop has finished and before another is planted – but they can equally be sown at various times of the year to help out with feeding when grazing is tight. They can provide a good source of food, extending the grazing season after grass growth has waned, as well as doing a few other useful jobs like breaking up the soil and reducing levels of pests and disease.

Most forage root crops are brassicas. Stubble turnips are one of the most popular and cheapest to grow. Sow just 2kg of seed to an acre and, within 10 to 12 weeks, you can expect 20 tonnes of forage. That would feed more than 30 sheep for a month, without any additional feed – and all for a bag of seed costing as little as £5 to £10, depending where you shop and how much you buy.

Although traditionally planted after a harvest, they can be sown into ploughed ground in late spring, once the risk of frost has passed. Sheep are normally allowed to graze the crop while it is still growing, as they eat both the leaves and the roots. Autumn sowing is another option, to give a crop that will last until January or beyond. For more about feeding root crops, see Chapter 8.

BUYING YOUR SHEEP

'Horses' for courses!

Just as gardening experts say it's best to match your plants to their environment, the same principle can roughly be applied to livestock. Of course, you *can* grow acid-loving azaleas in alkaline soil – as long as you're prepared to invest in a lot of ericacious compost. Similarly, you could plant drought-resistant species like lavender in a bed of heavy, wet clay – but while they may manage to survive, they won't flourish as they would in free-draining soil.

Switching back from flora to fauna, animals do best in those environments to which they have adapted, often over centuries. Sheep farming in the UK is described as 'stratified', which means that certain breeds tend to be suited to particular situations – in simple terms, higher or lower ground.

The general advice is that you should choose a breed to suit your terrain. That should make perfect sense: a breed that has been reared successfully in your particular area for generations is tried and tested in that particular environment, and therefore can be relied upon to do well. However, it doesn't mean that you can't ever rear a breed that wasn't specifically bred to suit your area; what it means is that you'll have to manage it carefully to ensure the best possible performance. Hill and upland breeds have adapted to make the best of poor grazing, so putting them on lush lowland grass for too long can result in some very fat sheep; on the other hand, some less resilient and less resourceful lowland breeds – including many of the continental commercial types – would not do well in harsh upland areas without considerable amounts of supplementary feeding and shelter from the elements.

If you set your heart on a rare breed, or one that isn't already reared locally, be aware that expanding or improving your flock could be difficult. Getting hold of a new, unrelated ram might not be easy, and you could end up clocking up hundreds of miles for the right one.

Breed characteristics, such as meatiness, appearance and ease of lambing, are all considerations when choosing a breed, but size, temperament and ease of handling should also be considered. If you choose a breed that is too big for you to catch and immobilise on your own, you will always have to call in a helper when you need to do a job, which means husbandry tasks may be delayed. Bear in mind, too, that while horned breeds may look impressive, they can make handling more difficult, and they can also be troublesome, getting caught in stock fencing, injuring other sheep, or growing abnormally, sometimes into the head.

At the end of the day, your own personal preferences are likely to govern what you choose. When rearing sheep on a small scale, there is no reason why they shouldn't be dual-purpose: useful but easy on the eye as well!

The range of breeds available in the UK is phenomenal – not just our own native breeds, but continental ones, too – so you really are going to be spoiled for choice. The guide in Appendix 1 has plenty of ideas for you, with more than 80 listed.

First, though, as touched on in Chapter 2, you have to decide why you want sheep and what you are going to do with them. For instance, are you interested in raising for meat, wool, milk, or other by-products? Each activity will have a bearing on what you choose. Whether you want to breed (and many people don't, which is absolutely fine) is another important consideration. Even if you just want a few friendly pets, there are still some basic guidelines to follow.

BREED CATEGORIES

Sheep breeds are divided roughly into categories, which denote their preferred environment or general purpose. They are sorted into groups – or 'stratified' – according to their natural environment, making them hill, upland, or lowland breeds. In addition, there are some hill and upland breeds that are classed as 'primitive'. See the breeds guide in Appendix 1 for details.

Hill breeds

Examples: Herdwick, Swaledale, Welsh Mountain, Scottish Blackface	■ Hardy and able to withstand extremes of weather. ■ Do well on poor vegetation. ■ Good browsers, so excellent for control of overgrown scrub. ■ Ewes are often crossed with upland rams to produce 'mules', which are crossed again with a purebred lowland breed to produce a commercial carcass. ■ Rarely need help lambing; good mothers.

SWALEDALE

Magali Pettifer

Upland breeds

Examples: Kerry Hill, Clun, Hill Radnor, Beulah	■ Hardy, but not as resilient as primitive or hill breeds. ■ Will survive on poor vegetation and unimproved grassland, but may need supplementary feed to maintain condition. ■ Rarely need help lambing; reasonable mothers.

CLUN

Court Llacca Cluns

Lowland breeds

Examples: Hampshire Down, Lincoln Longwool, Shropshire, Suffolk	■ Not as hardy. ■ Heavy breeds to handle. ■ Need improved grass to put on condition; must receive additional feed through winter. ■ Not good browsers. ■ Rams are popular choices as terminal sires to produce a meatier carcass. ■ Can require assistance during lambing.

LINCOLN LONGWOOL

Primitive

Examples: Boreray, Shetland, Soay, Hebridean, Castlemilk Moorit	■ Hardy and able to withstand extremes of weather. ■ Small and light-footed – often chosen for conservation grazing projects in sensitive environments. ■ Agile, active and good at escaping; can be nervous, making catching tricky, but light to handle once caught. ■ Do well on poor vegetation. ■ Good browsers, so excellent for control of scrub. ■ Too small and slow-growing for the commercial market, but sometimes crossed with lowland breeds; some niche market opportunities. ■ Easy lambers and very protective mothers.

CASTLEMILK MOORIT

Castlemilk Moorit Society

Where and how to buy

Knowing what you're looking for in an animal is your starting point. If you don't feel up to the task, ask someone more experienced to give you a hand.

It's easy to be tempted by adverts on the Internet for livestock, but when you're just starting off you should make life easy for yourself and buy healthy stock from reputable breeders who know what they're talking about. Good breeders will sell you decent stock: they're not going to risk their reputation by offloading something with health problems. Never be tempted to take on sheep – or, indeed, any livestock – if they are advertised as 'looking for a forever home'. Read between the lines and you may see the words 'high-maintenance geriatrics'!

From a biosecurity point of view, it's much safer to buy from a reputable farm in a one-to-one transaction than from a market filled with hundreds of sheep from numerous parts of the country. Reduce the risk of taking home disease if you possibly can.

If you don't mind what breed you have to start with, it might be worth finding out what your farming neighbours prefer, as anyone with a decent number of sheep will have times when they sell off surplus stock. Another plus point is that you will know that, if the sheep are already doing well on neighbouring farms, they will be suited to the environment.

If you are looking for a specific breed, start with the relevant breed society, which will have contact details for members. Visit agricultural shows and talk to exhibitors to see when they might have stock available. Don't be put off by the rosettes – only the cream of the crop make the show team and command massive price tags when sold. Every good breeder will have stock to sell at certain times of the year, and at a range of prices, depending on the standard.

FIXTURES AND FITTINGS

If you are just in the process of buying your farm or smallholding, you may find the vendor wants to offload machines, tools and even livestock. Don't be too ready to accept these 'bargains', unless you know for certain what you'll be buying. Are the sheep fit and healthy? Are they fit to breed from? How old are they? Are they really worth the asking price? Don't automatically assume you'll be getting a bargain. Take your time, settle into your new surroundings, and then start doing your homework – otherwise you could end up making a costly and disappointing mistake.

Buying at auction

Be extremely wary of buying at auction. It's easy to get carried away when bidding starts, and you may end up buying something you don't really want or need. What's more, if you get locked into a bidding war with another enthusiastic buyer, you could end up paying far more than you might have done had you bought direct from a farmer. Check current market prices before venturing into an auction – just as you would if buying a car. It's important to know what to look for in an animal before starting bidding, and with sheep the best way is by getting into the pen before the sale and doing a full examination. See pages 40–43 for advice on how to do a sheep 'MOT'.

Auctions: top tips

1 On your first visit to a livestock market, don't go with the intention of buying. Most large livestock centres have either weekly or monthly sales, so there's no need to rush. Go and do a recce first. Treat your first visit as an information-gathering exercise: see how the market works, what the procedure is for registering and getting your bidder's number; how much VAT and commission you'll have to pay on top; the methods of payment accepted; where and when you can load what you've bought into your trailer.

2 When you do pay another visit, with money at the ready, make sure that you have somewhere to put your sheep when you get them home. They should be offloaded as

Above: Arrive early to get a good spot near the sale ring.

soon as you arrive back – and not left cooped up for hours in a trailer while you get things ready.

3 Decide in advance how you'll transport your animals home. Make sure your trailer or vehicle is fit for purpose, and that it's roadworthy, clean and disinfected. If you're travelling a long distance, make sure you have the necessary transport permits in order (see the regulations in Chapter 2).

4 In advance of the auction contact the organisers for a copy of the catalogue. You may find it on their website. This will give you time to see what's on offer, do some research into breeds, and check the current market prices, so you don't bid over the odds. Most large operations post reports of previous sales on their websites. These normally include the prices of the top-selling stock and the vendors' names.

5 Take someone with you who knows their stuff. You can't beat the advice of someone who has experience of buying and rearing sheep.

6 Arrive early, register at the office to get your bidder's number, take a good look around, and note down the numbers of sheep (the 'lots') in which you're interested. If you arrive just when the auction is due to start, you won't have had time to assess the animals properly and you'll also be right at the back of the crowd, struggling to see what's being sold.

7 Make sure you know how many animals are being sold in each lot. If there is a pen of six sheep, bidding may be per animal or it could be for the entire group. Make sure you listen to what the auctioneer says before bidding starts.

Golden rules for selecting healthy stock

The same guidelines apply whatever species you are considering buying. Never buy anything that looks too quiet, or 'under the weather', hoping it will improve when you get it home. Go for good overall conformation – a strong, straight back, a deep body, to accommodate all the vital organs, a balanced body, and sound legs and feet.

Look out for:

- General appearance and posture. Disregard if it looks listless, has dull eyes, a droopy head, dry muzzle, or is reluctant to stand
- Discharge from eyes or nose
- Scouring (diarrhoea)
- Poor, dull or scruffy fleece/skin; bald spots or scabs
- Signs of itchiness – repeated rubbing against the pen or biting the fleece
- Coughing or sneezing
- Lameness or other signs suggesting pain or discomfort
- Lumps in udders (possibly mastitis)
- Worn, misaligned, or missing teeth, which could cause feeding problems

LAMB

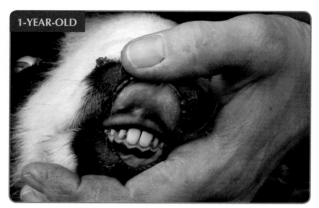

1-YEAR-OLD

2-YEAR-OLD

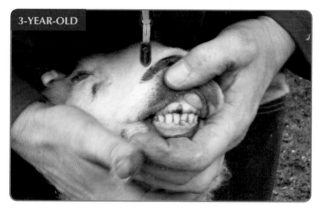

3-YEAR-OLD

Teeth, toes, teats – and testicles!

These are the key things you need to check when buying sheep. Remember them by thinking of them as the 'Three Ts'. Depending on whether you're selecting ewes or rams, substitute 'teats' or 'testicles' as appropriate!

Below: All the adult, permanent incisors are present, indicating a 'full-mouthed' sheep.

Central Incisors (I₁)

Incisor (I₃)

Incisor (I₂)

Canine or corner incisor (I₄)

Lower Jaw

The importance of teeth

Teeth are the best indicator of the age of a sheep, and also provide a guide to how well it will be able to feed itself. A mature sheep will have 38 teeth. Both the top and the bottom jaw will have premolars and molars towards the back of the mouth, which are for grinding food. The bottom jaw has a set of eight incisors – sharp cutting teeth that make contact with the (toothless) hard pad of the top jaw. These incisors are used to gauge the age of a sheep.

Within a few weeks of birth, lambs will have a full set of eight deciduous teeth – teeth that will eventually fall out as the adult incisors push through. Any time between the ages of 12 months and 15 months old the first two permanent incisors will erupt centrally through the gum at the front of the mouth. The other pairs of incisors follow at roughly six-month intervals – erupting either side of each previous pair – until all eight adult teeth are in place.

The incisors will eventually loosen and wear down, and may even break. An older sheep that has started to lose its teeth is known as a 'broken-mouth' or a 'broker'. Tooth loss can be a significant problem in commercial flocks, where good performance is vital, as it makes biting short vegetation and surviving on rough grazing more difficult and can lead to malnutrition and loss of condition. Such animals may be sold for culling before they start to lose too much weight, but often hill and upland ewes are sold on to farmers in lowland areas as 'draft' ewes. In less challenging surroundings, they can be as useful as younger ewes, provided there is plenty of grass of an appropriate height, or concentrate feed is given. Cared for in this way, they have the potential to carry on lambing for many more years, producing four or more crops of lambs. Buying in 'culls' can be a cost-effective way of building up a flock, but you have to keep in mind that you are buying to produce replacements. Eventually, the mothers will have to go, while their offspring take their place.

Under-shot and over-shot jaws

As well as checking to see how many incisors a sheep has, it's also important to make sure that they are correctly aligned with the upper pad, so that it can chew effectively. To check alignment, conduct a simple examination by holding the sheep's head in the normal position, keeping the mouth closed, and running an index finger along the dental pad. The main abnormalities to check for are under-shot jaw (brachygnathia) and over-shot jaw (prognathia). An under-shot jaw – sometimes known as a 'parrot mouth' – simply doesn't reach out far enough to meet the pad. An over-shot jaw ('monkey mouth') has the opposite characteristic, jutting out too far.

Both traits can be inherited, so any sheep found with these defects should not be used for breeding.

Check the cheeks

Look out for external swellings in the cheeks, which may be caused by dentigerous cysts in the rear of the mouth. These can cause significant problems when sheep are attempting to graze.

Worn or missing 'cheek' teeth – the molars and premolars – can pose a greater problem than lost incisors, because of the key part they play in grinding down fibrous matter. As well as external swelling and loss of condition, sheep with molar teeth problems may be seen with bits of grass or hay sticking out from the corners of the mouth, or staining where cud has

spilled through. They may also have problems eating concentrate feeds, often dropping cud and pellets as they chew. Run an index finger along the side of the face and feel for a raised, irregular ridge. The outer edges of the upper cheek teeth can develop sharp, jagged enamel surfaces, which can cut into the skin of the cheeks, allowing infection to set in and making examination very painful.

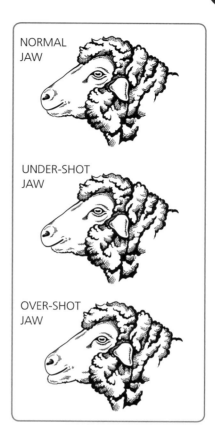

NORMAL JAW

UNDER-SHOT JAW

OVER-SHOT JAW

Feet and legs

A sheep with uncomfortable feet cannot and will not perform well. Painful feet will deter grazing and eventually lead to loss of condition, and the sheep won't be much in the mood for breeding, either. The structure of the legs is just as important, and you certainly shouldn't be breeding from anything that is less than correct (see illustrations). Chapter 6 looks at the issue of foot trimming – which has

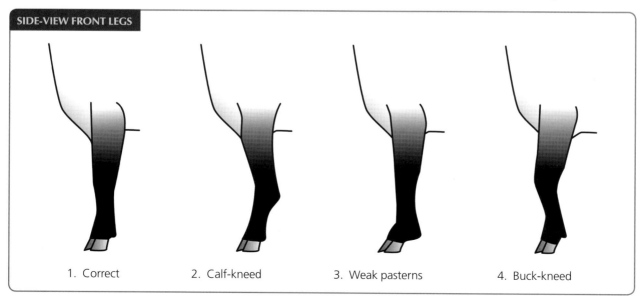

SIDE-VIEW FRONT LEGS

1. Correct   2. Calf-kneed   3. Weak pasterns   4. Buck-kneed

SIDE-VIEW REAR LEGS

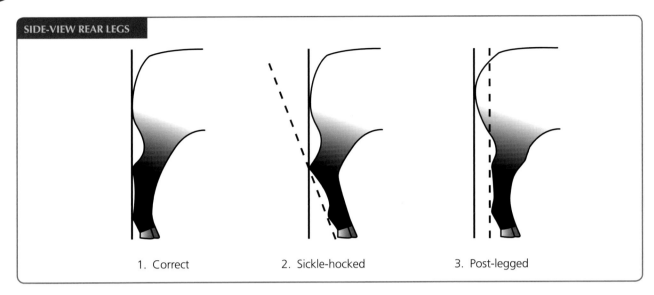

1. Correct   2. Sickle-hocked   3. Post-legged

FRONT VIEW

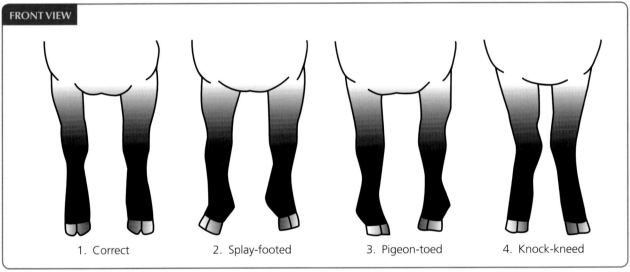

1. Correct   2. Splay-footed   3. Pigeon-toed   4. Knock-kneed

REAR VIEW

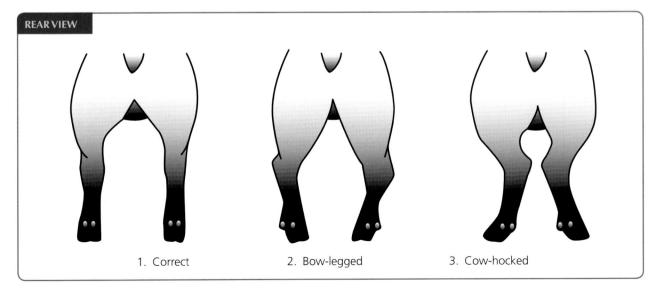

1. Correct   2. Bow-legged   3. Cow-hocked

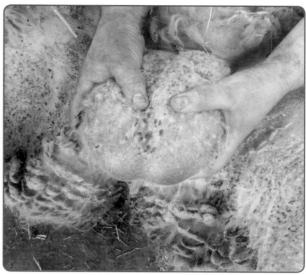

Jan Walton

Above: Feeling for abnormalities.

Below: Measuring the scrotum.

recently become quite a contentious issue – and the differing opinions on how much is actually needed. For now, though, let's concentrate on choosing a sheep that stands well.

A sheep should have sound feet and stand squarely with what is often described as 'a leg at each corner' – a bit like a table – so that it can move freely and comfortably, forage for food efficiently, and mate without problems. At sheep sales, pens are often too small to allow potential buyers the chance to see the animals moving normally. If you arrive sufficiently early, it's a good idea to ask the seller to lead the sheep out of the pen, so you can observe it walking.

Inspecting udders

Lumpy udders should be avoided as they can indicate chronic mastitis (more on this in Chapter 6). Mastitis is an inflammation of the udder and can be caused by physical injury, stress, or bacteria entering via the teat canal. It can impair colostrum and milk production, so lamb growth can be compromised.

Scrutinising the scrotum

Just as ewes get their undercarriage inspected, the important bits of rams need a good check-over, too. Gently feel the scrotum, seeking out the testes. They should be evenly sized and firm and springy to the touch – like a 'flexed bicep' is the general description! There should be no lumps or bumps. What you are feeling for is any abnormality in the spermatic cords and/or testes.

It's best to get someone to hold the ram in a standing or sitting position while you place your hands on the base of the scrotum, one each side. Scrotal size has been shown to have a bearing on the fertility of the ram in terms of semen production. Some research has shown that the size of the sire's testes might also affect the reproductive performance of female offspring. Generally, the scrotal circumference of a mature ram should be at least 32cm, while the measurement of a ram lamb should be no less than 30cm.

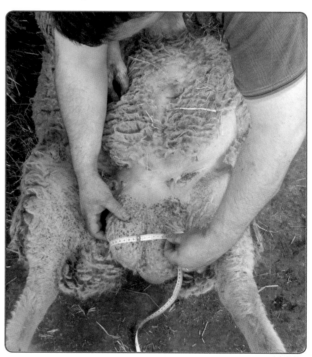

Jan Walton

Assessing overall condition

Sheep should have a good covering of flesh at all stages of life, but they should never be too fat nor too thin. There is a standard system of body condition scoring that is used for evaluating whether an animal is fit for breeding, and also for assessing whether it is ready to be killed for meat. See Chapter 8 for a full explanation of how this works. If you're unsure what to look for, ask a neighbouring farmer or your vet to show you how it's done.

BEHAVIOUR AND HANDLING

Sheep are definitely not stupid creatures. Their instinct to follow one another when panicked, seemingly regardless of whether it's the right thing to do or not, often leads people to brand them as dim-witted and gullible. Anyone who has tried, unsuccessfully, to herd sheep will testify otherwise.

The reason most sheep – with a few exceptions – run as one when they feel threatened is because they have no other way of protecting themselves from predators; as they can't fight back, they opt for the 'safety in numbers' approach, sticking together because it's far more difficult to be singled out as a target if you're part of a large group.

You'll notice that the flocking behaviour kicks in soon after birth, with lambs being taught by their mothers to follow. You will hear the ewes gently murmuring to their offspring to encourage them to stay close. Above all, though, sheep love the company of other sheep! They are incredibly sociable animals and need to be with their own kind far more than probably any other species of livestock. You only need to see how distressed and unsettled a sheep becomes when separated from the flock – either by accident or because it has to be separated for veterinary care – to understand how important it is for them to be part of a flock. In a stressful situation, such as when they are being herded into an enclosure, they will often find a corner and pack tightly together, because close body contact has a calming effect.

The flocking instinct can work to the shepherd's advantage in some cases, making it easier to move a large group in one go. Sheep are naturally docile creatures, and don't like stress or changes to their environment, so the best way of dealing with them is calmly and without yelling and shouting. Sheep have good memories; these memories need to be positive ones as much as possible. Once you understand their reactions and can anticipate their moves, you'll become far more proficient at handling them. If you want to understand how to manage sheep, you have to understand how their minds work.

Kay Hutchinson

Use of the senses

As a prime source of prey for many predators, sheep have learned to use their senses extremely well in order to keep out of harm's way. A sheep's hearing is highly developed and it's an important safety device, but eyesight is even more crucial to survival. Sheep depend heavily upon their vision and, in fact, are blessed with eyes designed to give them a wide field of view. While the human eye is forward facing and can cover an area of about 180 degrees, sheep eyes are placed more to the sides of the head, which means they can scan the surrounding area with only the need for slight movement. Depending on breed – and on how much wool they have covering their eyes – they are able to take in a view of up to 300 degrees, allowing them to perceive threats over large distances. Even when they have their heads down and are busily grazing, they can still see in all directions.

Having said that, they have little or no depth perception so, unlike us, they can't see in 3D; they therefore have problems identifying detail, which is why, when you want them to run through a gap you've created, they often won't see it properly. They never move easily into any situation where they can't see sufficiently well, which is why they will be reluctant to move from bright daylight into a darkened space, such as a barn or a trailer.

WORKING WITH THE FLIGHT ZONE

Every animal has its 'flight zone' – a certain amount of personal space in which it feels relaxed, comfortable and free from threat. Think of it as a kind of psychological buffer zone. Stay outside the flight zone and the animal will remain calm; step too far into it and it will be spooked, will instinctively panic and may injure itself, others around it, or the handler as it tries to escape. Sheep may not be naturally aggressive, but a fully-grown one – particularly one with horns – can do some serious damage when alarmed.

Hand-reared sheep ('pet lambs') may have no flight zone at all, feeling comfortable with humans at all times. This can make them useful animals to have in your flock, as their apparent ease in your company can sometimes persuade the others that you pose no risk. Conversely, sheep kept in flocks of hundreds or maybe thousands, which may be grazing enormous tracts of land and which rarely see a human, are more likely to have very large flight zones. Managing such a flock without the use of dogs and/or several helpers on quad bikes will be no mean feat.

Below: Pet lambs are easy to handle.

Left and above: Gathering pen for large numbers.

Below: Small home-made handling system.

Catching sheep

Everyone keeping sheep will need to round them up from time to time to carry out basic husbandry tasks or to load them into a trailer to go for sale or slaughter. Few people who are new to sheep keeping have the luxury of a purpose-made sheep-handling system – known as a 'race' – but it is possible to create your own handling area relatively cheaply, using hurdles, wooden pallets and other spare materials.

The ideal holding pen is big enough to contain all your sheep comfortably, without giving them too much space to move around or launch themselves over the sides. The smaller it is, the easier it should be to catch and restrain them, but you have to remember to leave yourself enough space to carry out any essential jobs, such as foot-trimming or shearing. The more efficient your handling system, the less stressful the experience will be – for the sheep and for you.

The ideal set-up is an initial 'gathering pen' where the sheep can be kept until you are ready to deal with them, an adjacent race or 'chute' (a narrow channel, just a bit wider than the sheep, designed so that they can't turn around once inside) with gates either end, and another gathering pen on the other side.

The race is such a useful piece of kit, making tasks like vaccination, drenching, the application of insecticides, tagging, or condition scoring easier for the handler, who can stand outside and reach over. The gates at either end of the race would ideally be the 'guillotine'-style, which open vertically to save space and reduce disruption. Another holding pen at the other end of the race is useful so that sheep which need to be handled in a larger area can be dealt with easily – such as at shearing time.

Pen Rashbass

Taking control

The easiest way to do anything to a sheep is to immobilise it. Approach it from behind, as its field of vision does not stretch that far, and work quickly and quietly. Decide which sheep you want to catch and wait for it to turn away from you – it probably will because your presence will make it and the others uneasy and their natural instincts will kick in. Make sure you have sufficient space behind you in order to manoeuvre the sheep into position – because you don't want to find yourself backed up against a wall or a fence with nowhere to go. If you can get one side of the sheep against a wall or fence before you start, even better. What you don't want to do at any point is to have any space between you and the sheep, otherwise it will sense the opportunity to escape.

The pictures and instructions here are for right-handed people, so just reverse things if you're left-handed. Some right-handed people prefer to do it the opposite way round, so find a way that works for you!

1 Firmly take hold of the sheep's head by cupping your right hand under its jaw and turning it into the left-hand side of its body. Don't worry about hurting the sheep by turning its neck round in this way, as sheep are amazingly flexible and can turn around to bite their own bottoms if they want to. At the same time as you are controlling the head, place your left hand on the rump to steady the sheep and press down into the flank.

2 Your knees should be behind the sheep, preventing it from moving. Remember to keep your legs together, or the sheep will slip through!

3 When you're confident you have control, take a step or two backwards, continuing to fold the head further into the body. The sheep should reach its natural 'tipping point', allowing you to take it down and then position it on its rump.

www.sheepeasy.co.uk

Above: If you don't think you can hold and treat a sheep at the same time, a sheep seat or a turnover crate (below) may help.

4 Once in this position, you should be able to restrain the sheep using a minimum of force, holding it in a relaxed position between your legs. With practice, you'll find you don't even need to use your hands. What you don't want to do is to allow the sheep to have its hind legs making contact with the floor. If it does, it will try to push against the ground for leverage and try to escape.

Sheep held in a sufficiently small space can be caught by stretching under the belly and grasping the hind leg on the opposite side, and then taking hold of the jaw as before. This approach is best if there are two people. Lambs and small breeds can be caught by putting your arms around the shoulders, taking hold of the front legs, lifting them, and pushing the rump forward with your knee at the same time.

A sheep seat is a handy device to have, particularly if you suffer with a bad back, or if you find that keeping control while carrying out husbandry tasks at the same time is too much for you. It's basically a kind of extra-strong deckchair in which your sheep sits, replicating the position it would be in if you were holding it between your legs. The top part of the seat hooks over the top of a gate or fence. Once reclined,

the sheep shouldn't put up too much resistance, but some people prefer to add a belt around the seat and the sheep's belly in order to give additional peace of mind.

A more sophisticated – and much more expensive – idea is the turnover crate. This is a metal crate that the sheep is put into and then either a lever is pulled or a wheel turned to rotate it safely.

WARNING

Never attempt to catch a sheep by grabbing hold of its wool. As well as hurting the sheep, it can cause bruising to the carcass, which takes a long time to subside and which may still be visible at slaughter time.

Handling rams

Always be cautious around rams, particularly during the breeding season. Even the most docile, hand-reared rams can change temperament when the hormones start working and can become very aggressive, challenging and headbutting fellow sheep, humans and inanimate objects like walls and gates. Always be cautious and never allow children to be around them unsupervised – whatever the time of year.

Do you need a sheepdog?

If you only have a small number of sheep, the answer is probably no. If you train your sheep to recognise the sound of food in a bucket, you should be able to get them to follow you with ease. With larger flocks ranging over

considerable areas of land, a properly trained dog in the hands of a competent handler can do the job of several people. But it's not simply a case of buying any old dog and letting it loose on your sheep; good sheepdogs are carefully bred for both aptitude and temperament and can take several years to train. Both dog and handler have to be trained, of course, and there are plenty of organisations and individuals offering tuition in this complex art.

When you're confident you can handle a dog correctly, you may want to buy an experienced dog with which to work. Two novices aren't going to achieve much working together, so at least if one of you has something of a track record, you'll improve your chances of success. Aim for one registered with the International Sheepdog Society (ISDS), but don't expect much change out of a few thousand pounds for a good one. You may get a part-trained dog needing more work for a few hundred, but, as with most things in life, you get what you pay for.

Popular shepherding breeds

In the UK, the Border Collie (right) is often seen as the archetypal sheepdog. Although they come in several colours, black and white – or black, white and tan – remains the most popular. As the ancient Celts were thought to be the first to use dogs for shepherding, it's often thought that the name 'collie' comes from the old Welsh 'gellgi', which meant 'covert hound'.

There are two basic types of 'top notch' Border Collie – those registered by the ISDS, which are admitted into the stud book on skill and ability, and those registered by the pedigree organisation, the Kennel Club. The latter have been bred more for appearance than working characteristics. There are, of course, many more working dogs that are not registered by either organisation – and which therefore can't accurately be described as a 'Border Collie', even though they may look like one.

The Welsh Sheepdog is often confused with the Border Collie as, although the most regularly seen coat colour is red and white, they can, like the Border Collie, vary in appearance. The Welsh Sheepdog Society was set up in 1997 to preserve the purity of the breed and promote its benefits as a working animal. Although keen to see the breed grow in popularity, the society makes no secret of the fact that it would prefer to see registered puppies being sold solely to working farms.

Shaggy-coated Bearded Collies were originally used as droving dogs and are often used on farms for managing cattle as well as sheep. Slightly higher-maintenance, as they have to be clipped, they are sometimes described as easier to handle and more adaptable than the traditional shepherding breeds.

More and more farmers are looking to the Australian Kelpie as an alternative herding dog. Their appearance suggests links to the Border Collie, and this seems likely as many were exported to Australia early last century. The breed has all the instincts of a collie, but may be a little too strong-willed for an inexperienced handler.

Strong and powerful, this long-legged dog from New Zealand is a barker rather than a biter, without the 'nipping' instinct of the collie.

Praised for its good temperament,

WELSH SHEEPDOG

BEARDED COLLIE

the Huntaway is becoming more and more popular in European countries. Although it's likely the collie played a part in producing the breed, the size, colouring and demeanour suggests there may also be a bit of German Shepherd, Rottweiler, or even Labrador in the background.

KELPIE

HUNTAWAY

Oogie McGuire

Flock-guarding dogs

Dogs aren't just used for rounding up sheep; they are also used in many countries to protect the flock from predators such as bears, coyotes and mountain lions. Popular breeds include the Maremma, Spanish Mastiff, Pyrenean Mountain Dog and Polish Tatra. These dogs are brought up with the flocks as puppies and live to defend them.

SPANISH MASTIFF

POLISH TATRA

Ricky Smyth

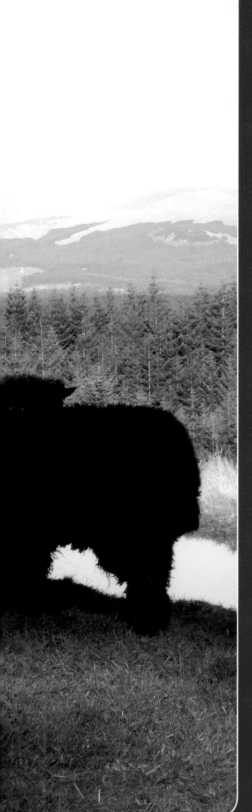

CHAPTER 6 SHEEP MANUAL

A HEALTHY FLOCK

It's not unusual to hear experienced old farmers saying, 'Sheep are always looking for new ways to die.' Don't let that put you off! Human negligence and lack of vigilance is often to blame for common health problems affecting sheep. Sheep can be pretty high-maintenance animals, and the phrase 'prevention is better than cure' is highly appropriate in their case. If you can take preventative measures to safeguard your flock from some of the more common problems, you'll save yourself a lot of work, heartache and expense in the long run.

One of the best advantages you can give yourself is to start with healthy stock bought from a reputable source, so refer back to Chapter 4 for advice on how to choose the best. Start with bad stock and you'll make life much harder for yourself.

Observing your sheep is key to nipping ailments in the bud. If you get to know your flock well and keep an eye on it on a regular basis, you'll soon be able to spot telltale signs when something is wrong. You need to know what counts as normal behaviour before you can detect what's abnormal, so spend time watching your sheep whenever you can. It will be time well spent.

Vaccinations to consider

Have a chat to your vet about whether or not you should vaccinate your flock against some of the most frequently seen diseases. Remember that any veterinary products you use must be properly recorded and the withdrawal periods observed – as explained in Chapter 2.

Clostridial diseases

There are various different types of disease caused by clostridial bacteria. Broadly speaking, they are divided into three groups, which are:

- Diseases affecting the digestive tract and internal organs
  Abomastitis
  Bacillary haemoglobinuria
  Black disease
  Braxy
  Lamb dysentery
  Pulpy kidney
  Struck
  Toxaemia
- Those that cause muscle damage or gangrene and toxins in the blood
  Big head
  Blackleg
  Gangrenous metritis and joint/navel ill
  Malignant oedema
- Those that cause nervous damage
  Botulism
  Focal symmetrical encephalomalacia

Spores of clostridial bacteria can exist in the soil for many years, but healthy, content sheep are not always affected by them. Sheep become vulnerable when stressed by a change of diet or environment, if injuries cause open wounds, or if

FIND YOURSELF A VET

Even before you get your first sheep, you should be finding out where the nearest farm vet is based. Don't wait until you need a vet for a sick animal – vets who treat livestock can be hard to find in some areas, and you may have to look outside your immediate area and possibly pay an additional call-out charge to cover the travelling involved.

Most vets will want to visit your stock soon after you register. Use such a visit to have your flock checked over and ask the vet to advise you of routine tasks to be carried out, such as worming and vaccination. Every area will have its different disease concerns, and a vet who is familiar with neighbouring farms and their stock will be best placed to offer advice on any preventative treatments to give. He or she will also be happy to show you how to administer drugs and explain, for instance, the sites for different types of injections.

If you're just starting off, do call out your vet for routine jobs initially, until you feel more confident. The time subsequent treatments are needed, you'll probably feel able to do the job yourself.

GIVING INJECTIONS

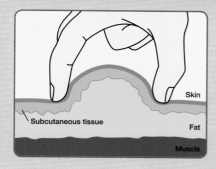

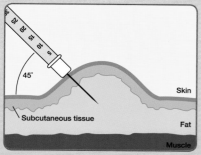

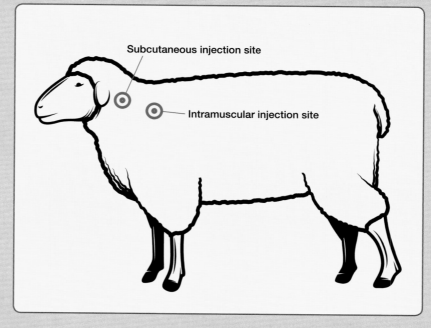

Most injections you will need to give will be administered subcutaneously – under the skin, for slow release. Relatively few sheep drugs require intramuscular injections – given directly into the muscle, with a needle held vertically – and those requiring an intravenous route are best left to your vet when you are just starting out.

Subcutaneous injections are most easily given if the skin is pinched into a 'tent' shape and the needle inserted into it at a 45-degree angle. Be careful not to let the needle protrude out the other side, though, otherwise the drug will be wasted. Also, you might inject your own fingers in the process!

Always check the length and gauge (thickness) of needle required for the drug and dispose of needles, syringes and used bottles appropriately. Record everything in your veterinary records book.

Below and right: Choose the correct size needle for the job and dispose of it carefully.

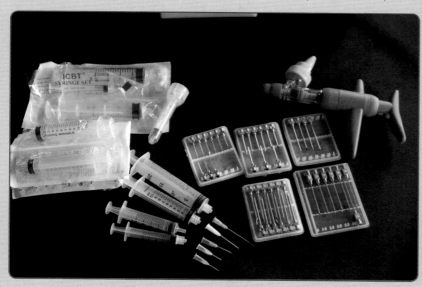

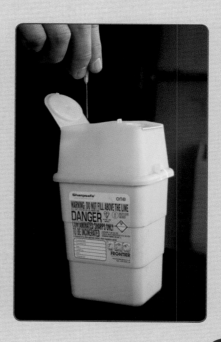

QUARANTINE

All new stock should be quarantined before being mixed with your existing flock. It's worth setting up a dedicated isolation unit to hold new purchases before you introduce them to your existing flock, and the same goes for sheep taken to agricultural shows or stud rams hired out. Contact your local Animal and Plant Health Agency office for advice (see Contacts on page 182).

they are affected by internal parasites. Symptoms may not be apparent, and sometimes the sheep simply drops dead before illness is detected.

Sheep that have not previously been vaccinated should be given two doses, four to six weeks apart, with annual boosters in subsequent years. Ewes should have their boosters four weeks before lambing, in order to pass on protective antibodies to lambs via the colostrum. Lambs born to ewes that have been vaccinated are considered to be resistant for up to 12 weeks, but they should be vaccinated for ongoing protection. The vaccine is normally given from about eight weeks old, with the second dose four to six weeks later. Lambs from unvaccinated ewes should be treated earlier – but check the information sheet supplied with the vaccine for minimum age restrictions, which can vary between manufacturers.

One drawback for small-scale sheep keepers is that once the vaccine has been opened, it must be used immediately or disposed of safely. If you are giving two separate doses, you will need a fresh bottle for the second vaccination. It's not an expensive vaccine, but your vet may be able to put you in touch with others in your area who are using the same product so that you can coordinate vaccination times and share the cost. Several veterinary practices are starting to recognise that not all pharmaceutical companies produce things in convenient sizes. It's a well-known fact that some smallholders are deterred from buying preventative products if they are available only in large quantities and at a significant price, so a good veterinary practice will be keen to encourage vaccine sharing where appropriate. Never keep the first bottle, thinking you can re-use it for the second dose. You'll be wasting your time and leaving your sheep unprotected.

Pasteurellosis

This is the most serious respiratory disease in sheep and is caused by Pasteurella bacteria. Most vaccines for clostridial disease will also protect against pasteurellosis (in Heptavac-P Plus, the 'P' stands for 'Pasteurella'), but there are some specifically designed to combat Pasteurella alone. Pasteurellosis can be treated with antibiotics, but vaccination is worth thinking about if you have a recurring problem.

Escherichia coli (E. coli)

This bacterium causes a variety of problems, one of the most common being 'watery mouth disease' in newborns. Lambs are listless, reluctant to get up and suckle, and salivate profusely – hence the name. Death can follow swiftly. Dirty, wet conditions in lambing sheds can raise the risk of E. coli infection. Ewes can be vaccinated against E. coli, with two doses given at least two weeks apart. The second dose should be administered four weeks before lambing, so that immunity is passed to the lambs via the colostrum.

Salmonella

It isn't normal practice to vaccinate against Salmonella, but your vet may suggest it if other species on the farm are known to be affected, or where ewes have aborted and the reason has been shown to be salmonellosis.

Orf

There are a number of zoonotic diseases – those that can be transferred from animals to humans – and orf is a contagious skin condition, which can be uncomfortable for both sheep and handler.

Careful handling of sheep on farms where orf exists is essential, so take strict hygiene precautions, wear gloves and, if you have cuts or abrasions, take care to protect them from potential infection.

Orf normally affects lambs up to a year old, but can be passed on to ewes. The orf virus is pretty resilient and can live in soil and in barns for several years. Sheep normally become infected when skin or gums are pierced, such as when

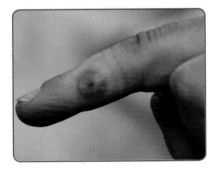

Above and below: Orf on a shepherd's finger (above) and on a lamb's mouth.

MAEDI VISNA (MV)

If you've visited the sheep sheds at agricultural shows, you may have noticed the sheep are not only divided by breed or type, but also whether or not they are 'Maedi-visna accredited'.

MV is a chronic disease of sheep caused by a retrovirus that can have a disastrous effect on production, with increased ewe mortality, lamb losses and poor growth rates. The name derives from two Icelandic words that describe the main clinical signs of pneumonia and wasting – 'maedi' refers to affecting the lungs, while 'visna' refers to affecting the central nervous system.

The disease has a long incubation period and signs can go unnoticed – sometimes not until the sheep is as much as three years old. Symptoms include pneumonia, progressive paralysis, arthritis, muscle-wasting and weight loss, loss of milk supply and chronic mastitis. There is no cure and no vaccine.

The Maedi Visna Accreditation Scheme was set up to combat the disease. It's a voluntary scheme and members have to agree to the rules and conditions of membership, which include undergoing regular testing. The idea behind the scheme is to breed solely from MV-free sheep and to assure buyers that, if they buy from an MV-accredited flock, they can be confident that they're not buying in the disease. Some agricultural shows will not allow sheep that are not accredited,

while others have separate sections for those that are accredited and those that are not.

There is also the Maedi Visna Monitoring Scheme – a completely separate scheme. A flock can become MV monitored when it passes one qualifying blood test. Monitored sheep are blood-tested every two years and a certificate is issued if the flock passes.

Sheep that are accredited must not have contact with any sheep that are not – including monitored sheep. Monitored sheep must not be exposed to non-accredited sheep or non-monitored sheep for more than one day.

grazing rough pasture containing thistles or gorse. The skin around the mouth and nose becomes red, with swollen pustules; this is a painful complaint, making suckling or grazing difficult and leading to loss of condition. Scabs appear, which can drop off, spreading the virus further. In severe conditions, the tongue, gums and the inside of the lips and cheeks can be affected.

Lambs can pass orf to the ewe's legs and udder during suckling, and it can cause mastitis. The ewe's discomfort may stop her feeding her lambs. In humans, orf most commonly affects the hands and arms, entering through skin lesions. The virus can be transmitted to other sheep in the herd through close contact when feeding at troughs.

Although debilitating for the sheep, orf normally takes between four and six weeks to run its course, and results in a full recovery. It can't be treated once the virus has taken hold, but it is worth using antibiotic sprays on affected areas to reduce the risk of bacterial infection.

There is an efficient vaccine available, but, as it contains live orf virus, it should only ever be used on land where orf is known to exist – never as a preventative. Lambs born to vaccinated ewes aren't able to obtain immunity, so they should be vaccinated soon after birth.

Coccidiosis

Caused by a small parasite in the intestines, this disease often affects lambs from four to about eight weeks old and is responsible for acute scouring (diarrhoea), sometimes containing blood, abdominal pain and lack of appetite. Dehydration follows, lambs lose condition and often die. It is more of a problem where sheep are raised intensively, as lambs ingest coccidial eggs (oocysts) from infected pasture and bedding. Lambs born indoors are also more likely to be affected.

There are several treatments as well as preventative drugs available, but flock management will also need to be improved, taking care to keep stocking densities down, observing good hygiene and having regular cleaning in lambing sheds, and avoiding grazing very young lambs with older ones.

Johne's Disease (Paratuberculosis)

A chronic infection of the intestines, this disease is spread by ruminants ingesting a bacterium that causes progressive weight loss and, eventually, death. Faeces from infected sheep carry the bacteria and it can also be transmitted via the placenta to foetuses and to newborns through milk. The long incubation time of two to four years and lack of symptoms mean this disease can be in a flock for some time before being diagnosed. Some sheep can be infected but show no signs, and can act as carriers. Vaccination is the only effective method of control.

Schmallenberg

The Schmallenberg virus (SBV) is an insect-born infection affecting sheep, goats and cattle, and causes stillbirths, abortion and deformities if females are bitten while pregnant. Offspring are typically born with malformed joints and twisted legs and can sometimes be so badly disabled they have to be delivered by caesarean section. Abnormal foetuses are more likely if ewes are infected between days 28 and 56 of pregnancy. Badly affected flocks have been known to lose between 25 and 50% of their lambs.

The disease was first discovered in Germany in 2011 and has, since then, spread across Europe, with the first outbreak in the UK in 2012. A vaccine is now available and is worthwhile for those who breed for early-season lamb, as ewes are more likely to be pregnant when the midges carrying the virus are around. Ewes that are impregnated later in the year to lamb after February the following year are less likely to be affected. The vaccine is administered as a single dose in sheep (cattle require two) and can be given from the age of 10 weeks. Immunity is achieved within three weeks.

Below: This lamb with Schmallenberg disease survived and is thriving, despite her deformities

Above: A Schmallenberg lamb

Rachel Graham

Lizzy Fenn

CHECKING VITAL SIGNS

If you ring your vet for advice, he or she will probably ask if you have checked the 'vital signs' – temperature, heartbeat and pulse.

- The normal rectal temperature of sheep is 39.5°C, but it can vary up or down by 0.5°C. In lambs, it should be between 39 and 40°C. Using a lubricated veterinary thermometer, restrain the sheep in a standing position, lift the tail, and gently insert the thermometer, rotating it slightly until it goes in to two-thirds of its length. Tilt it so that the bulb rests on the bowel wall. Hold it there for up to two minutes, and then read the temperature. If in doubt, check again. An increase in temperature may indicate an infection, inflammation, or toxins. Take into consideration environmental factors such as weather and building temperatures.

- Heartbeat in a mature sheep should be around 75 beats per minute, but can range from 60 to 120. Place a stethoscope on the left side of the chest, just behind and above the elbow where there is normally less wool. Count the number of beats per 15 seconds and multiply by four to give the heart rate for one minute.
- The pulse can be taken by feeling the femoral artery (inside the hind leg about a third of the way towards the back of the leg where the muscles meet the abdominal wall). Use very light touch with one finger. A normal pulse rate will be 60–80 beats per minute.
- Respiration rates can be difficult to assess but 12–20 breaths a minute is considered normal.

Notifiable diseases

By law, certain diseases have to be reported to your local Animal and Plant Health Agency (see page 182).

Bluetongue

This is another disease transmitted by midges and affecting ruminants. Animals develop flu-like symptoms, swelling and haemorrhaging in and around the mouth and nose, foot lesions and may become lame. The tongue can become discoloured – turning blue or purple – which gives the disease its name. Death can occur within a week of symptoms being spotted. A vaccination made available several years ago appears to have removed the problem in the UK, though other parts of the world, where different strains exist, are still affected.

Foot and mouth disease

Symptoms can vary from listlessness and slight lameness to characteristic blistering in the mouth and around the top of the feet. FMD is an infectious viral disease, spread by aerosol means (e.g. coughing or sneezing), and, although most animals would recover given time, it can be fatal in newborn livestock. Control is by containment and culling.

Scrapie

A fatal, degenerative neurological disease with similarities to BSE in cattle and Creutzfeldt-Jakob disease (vCJD) in humans. Early signs of infection include behaviour changes such as lack of co-ordination and unsteadiness, abnormal walking, repeated scratching against objects and trembling. There is no cure and no vaccine available.

Anyone wishing to export breeding sheep, semen, or embryos to other EU countries must be a member of the Scrapie Monitoring Scheme (SMS). Flocks need to be examined by a vet over three consecutive years for signs of the disease before being admitted to the scheme. There must be no mixing with unmonitored flocks – including at agricultural shows – and any new stock bought in must be from monitored flocks. Fields or buildings used must not have been occupied by unmonitored sheep in the preceding three years, and any sheep over 18 months old that are culled or that die must be submitted for testing.

Tuberculosis

Although not normally thought of as a disease of sheep, instances of Bovine Tuberculosis (bTB) have increased in recent years. Sheep can be infected with Mycobacterium bovis (M. bovis) but they are normally not identified until the carcass is examined at the abattoir. If lesions suspected to be bTB are discovered, the Food Standards Agency will arrange collection and analysis of samples.

Anthrax

To put it into perspective, the last case in the UK was in 2006. Although suffering from the disease, sheep found dead may show no obvious signs of the disease. Some may show signs of high temperature, fever, breathing problems, shivering or twitching, blood in dung or coming from the nostrils, loss of milk production, seizures, staring eyes, colic-like discomfort; general depression and loss of appetite.

The disease (being zoonotic in nature) can affect humans and can initially give flu-like symptoms. Direct contact can cause an uncomfortable skin infection, which can cause boils. Inhaled anthrax spores can cause lung damage, which can be fatal.

Brucellosis

Brucellosis is a highly contagious disease of livestock and is characterised by abortions or reproductive failure in sheep and other mammals. Although they may recover, they can continue to shed the bacteria, spreading the disease. The bacteria, which is present in birth fluids and other excreta, can remain active in the soil for some time.

Brucellosis is another zoonotic disease which causes flu-like symptoms and/or persistent headaches. Infection can enter the body through the eyes, inhalation, swallowing or through skin wounds. Infection can also occur through drinking unpasteurised milk from infected animals.

Rabies

This is a disease affecting all wool- or hair-covered mammals capable of producing milk – including humans and household pets. Although there have been no notifications in the UK for several decades, the border entry points still remain on alert.

As it is found in the saliva of infected animals, it is normally spread by the bite of an infected animal.

Early signs can include behaviour changes – particularly aggression or shyness, excessive attention seeking, hypersensitivity to noise or light, excessive salivation, itching and extreme thirst. As the condition worsens, muscle weakness sets in, particularly in the legs and tail, the animal may show difficulty swallowing, be frothing at the mouth, and the eyelids may be drooping. Paralysis follows, and then convulsions, coma and death.

Below: Foot and mouth disease can shut down the countryside.

Understanding feet and lameness

Problems with feet can be the bane of any shepherd's life. Lameness may be easier to spot than other health problems, but understanding the cause and knowing how to treat the condition – as well as reducing the likelihood of it happening in future – should be your aim.

Before being able to identify and deal with problems, you first need to understand a little about sheep feet and be able to identify what a healthy foot looks like.

The best way of getting to know what a healthy, normal-shaped foot should look like is to examine the feet of a lamb. The foot has two cleats or toes surrounded by walls of horn, with much softer pads underneath. A sheep's foot is designed so that the horn wall sits slightly proud of the pads and takes the weight, protecting the softer parts of the foot from potential damage. Depending on the kind of surfaces sheep are grazing, the horn can occasionally grow longer, curling under the sole. Old farming advice would have been to trim this away, but experiments with farmers who have stopped routinely trimming have observed fewer instances of lameness.

Below: The horn on this foot is starting to grow under, but does not need trimming at this stage.

WHEN TO TRIM?

For generations, farmers have been carrying out routine foot trimming on a regular basis in the belief that feet need trimming to prevent lameness occurring. Some may have carried out trimming as often as twice a year, or whenever the sheep were gathered for other husbandry tasks. For many years, many shepherds were taught that good, hard paring back of feet was essential – and if they bled in the process, that was a good thing!

Modern advice from veterinary experts is NOT to trim unless absolutely necessary. Research has shown that inappropriate trimming can not only damage the structure of the foot, but also cause serious problems and exacerbate conditions (see opposite). Studies have shown that trimming can also delay recovery from several infectious forms of lameness and increase the spread to others in the flock. It's now generally accepted that trimming should only be carried out in a few situations, such as improving the shape of a seriously overgrown or misshapen foot; to allow proper diagnosis and treatment of the cause of the lameness; to allow the cleaning of abscesses or to remove trapped dirt, stones, etc.

The author is grateful to Professor Laura Green and her team at the School of Life Sciences at the University of Warwick for providing the photographs of common foot problems. Professor Green heads a research group looking at ways of improving the health and welfare of farmed livestock. For more information, see http://www2.warwick.ac.uk/fac/sci/lifesci/research/greengroup

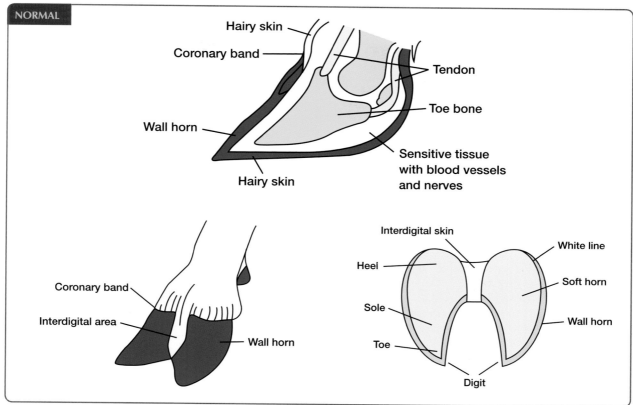

NORMAL

Hairy skin

Coronary band

Tendon

Toe bone

Wall horn

Hairy skin

Sensitive tissue with blood vessels and nerves

Coronary band

Interdigital area

Wall horn

Interdigital skin

Heel

White line

Sole

Soft horn

Toe

Wall horn

Digit

Common foot problems

There are numerous things that can cause a sheep to go lame, some of which are easily confused or misdiagnosed.

Scald (interdigital dermatitis)

This problem starts as a nasty irritation between the toes and is probably the most common cause of lameness in sheep. It typically occurs when ground conditions are wet or when grass is longer than normal, but it can occur when sheep are kept indoors and when bedding straw is wet and warm. The skin between the toes becomes red, moist, swollen and increasingly uncomfortable. Mild cases can be treated using oxytetracycline aerosol sprays, but more severe cases may need a long-acting antibiotic, too. Lambs affected should only be treated by spraying.

Foot bathing in zinc sulphate, copper sulphate, or formalin is also recommended, particularly when a number of the flock are affected. After walking them through the bath, the sheep must be allowed to stand on dry ground to allow the preparation to dry. Veterinary advice is NOT to trim.

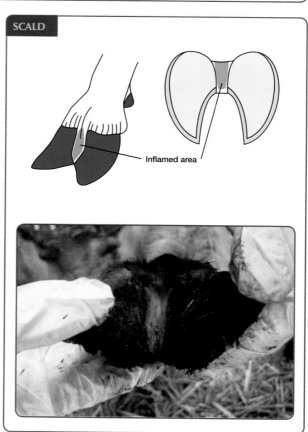

SCALD

Inflamed area

FOOT ROT

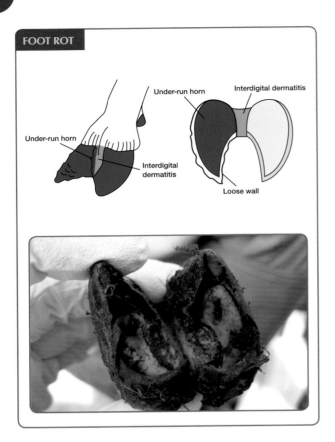

CODD

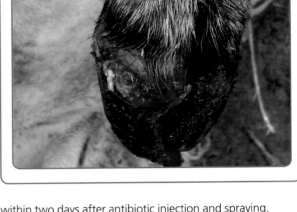

Foot rot

Foot rot is much more painful than scald and its effects are so severe that sheep are reluctant to move around and graze and so lose condition rapidly. If both front feet are affected, you may see sheep feeding on their knees.

The first physical signs are similar to scald, with the skin between the toes swelling and becoming moist, but infection spreads under the horn tissue causing the horn to loosen and separate. The walls of the hoof and the toes can become overgrown, trapping the infection underneath, along with dirt and other debris. The foot will have a characteristic unpleasant odour, which once smelled is never forgotten, and you may also see grey pus. An additional risk with foot rot is that the feet can be a target for blowflies when the sheep is lying down, so watch out for maggots.

Treatment methods include spraying, antibiotics and foot bathing in antibacterial solutions, but more and more farmers are attempting to breed out foot rot by culling susceptible sheep. Preventative vaccines, which are highly effective for four to six months, are also available for those with serious, recurring problems. Good husbandry, such as not overstocking pastures or sheds, isolating animals that are being treated and quarantining any bought-in stock, is always recommended.

As with scald, foot trimming is not recommended, as it has been shown to have no effect in preventing infection. Trimming has been proven to actually *delay* healing; research from the University of Warwick showed that 50% recover

within two days after antibiotic injection and spraying. Trimming delayed recovery by another eight days. Take advice from your vet before attempting any cutting back.

CODD (Contagious Ovine Digital Dermatitis)

This is a relatively new arrival, first seen in the UK in the late 1990s. Unlike the first two conditions mentioned, this is one that doesn't affect the skin between the toes.

Instead, an ulcerated area appears at the coronary band (the top of the hoof) and then the infection spreads under the horn and down to the toe. The horn detaches and falls off, and there is hair loss stretching to a few centimetres above the coronary band. Spraying the feet and treatment with antibiotic injection is recommended, with Tilmicosin being shown to have the best results. Again, don't trim feet.

Controlling this condition involves buying from unaffected flocks, quarantining new purchases for at least 30 days, and isolation of any lame sheep while they are being treated.

Toe granuloma

Most often caused by over-enthusiastic and ill-advised foot trimming, excessive use of formalin footbaths, or failing to treat foot rot soon enough. Sheep become severely lame and lose condition. A fleshy lump of granulation tissue – about the size of a pea and sometimes with a strawberry-like appearance – appears under the sole. The lump is extremely sensitive and bleeds easily. Advice is to apply a copper

TOE GRANULOMA

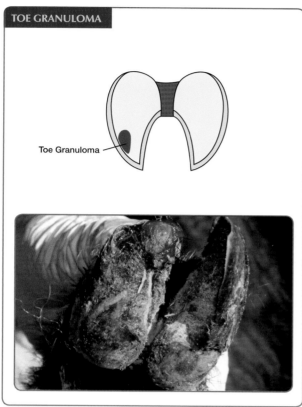

Toe Granuloma

WHITE LINE/FOOT ABSCESS

Pus 'pops' out at coronary band

Site of penetration (may not be visible)

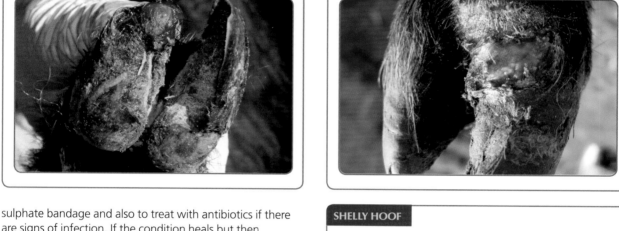

sulphate bandage and also to treat with antibiotics if there are signs of infection. If the condition heals but then reappears, culling is recommended.

White line/foot abscess

The 'white line' is where the wall horn joins the sole horn. This condition causes severe lameness and often occurs when bacteria enters through a lesion caused, for instance, by a thorn or other sharp object. This causes an abscess to develop under the horn and you may see pus emerging from the coronary band. The foot may feel hot and cause the sheep pain when touched.

This is a situation where careful paring back of the sole may need to be carried out, in order to drain the abscess and release pressure. Treat immediately with antibiotic injection and spray.

Shelly hoof

Often treated when there is no need, this condition sees the horn of the toe and the foot wall separating, creating a pocket for dirt and debris. Experts are divided as to the cause of shelly hoof, with some suggesting it could result from damage, wet ground, or a nutritional imbalance. It doesn't always cause lameness. Veterinary advice is that, if the sheep is lame and the horn is clearly separating, you should clean out any impacted debris and trim off any loose horn. However, if no lameness is seen, leave the sheep alone but monitor the condition.

SHELLY HOOF

Mud/stone

Loose wall horn removed

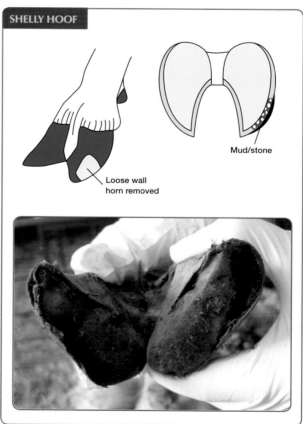

Unwanted visitors: dealing with parasites

Endoparasites

Internal worms pose a major health challenge to sheep, particularly to lambs, which are far more vulnerable. Worms can cause loss of appetite and permanent damage to the gut, they prevent nutrients being absorbed efficiently, reduce protein metabolism, and slow down muscle growth and overall carcass development. They can cause as much as a 50% reduction in growth rate – often without lambs showing any clinical signs.

Working to eradicate worm problems means knowing exactly what parasites you are dealing with and how severe a threat they pose. Over-use – and, indeed, misuse – of anthelmintics (worming treatments) over several decades has resulted in widespread resistance across the world. New anthelmintics have been developed in response to the growing resistance, but the advice is not to simply switch types, but to understand more about how the worm burden in your flock and on your land can be reduced.

The idea of treating 'just in case' is now regarded as outdated and modern thinking is to take samples of fresh faeces for analysis. Faecal egg counts (FEC) give an idea of the amount of worms in the gut and also identify the types present. You should consider carrying out one before dosing your sheep with wormer.

All vets will offer a FEC service and there are also companies online who will analyse samples for you. If you're scientifically-minded and have an old microscope from your student days, it's possible to carry out the test yourself.

Types of internal parasites

Most types of roundworm have similar life cycles (Nematodirus being an exception – see below). Ewes and ewe

Roundworm life cycle

Ingested larvae mature in intestinal tract

Eggs shed in faeces

1st stage larvae

2nd stage larvae

3rd stage larvae, only infective stage

SCOPS

Sustainable Control of Parasites in Sheep (SCOPS) is a group, led by key players in the sheep industry, that was set up to help find ways of improving parasite control and raise awareness of the need to use anthelmintics responsibly. Its website, www.scops.org.uk is packed with useful information for flock owners and, as well as holding a library of excellent publications, it posts timely reminders when high-risk times of the year are approaching.

The National Animal Disease Information Service (NADIS) also publishes a regular 'parasite forecast', using meteorological information to predict the prevalence of parasitic diseases at various times during the year. See the website www.nadis.org.uk for more information.

lambs build up an immunity during their first year of life, but lose resistance around lambing time and whilst suckling young. Rams tend not to build up effective immunity.

Nematodirus battus (causes nematodirosis)

This worm has a longer life cycle than most and eggs are prompted to hatch when very cold weather is followed by a temperature rise to about 10C or above. Larvae are ingested by lambs and the resulting nematodirosis causes severe scouring which is often dark, containing blood. Dehydration and general weakness follow, often resulting in death.

Lambs between one month and three months old are the most susceptible to nematodirosis, so if they are more than three months old when the weather starts to warm up, they may have escaped the most vulnerable stage. Similarly, if mild weather arrives early, triggering the eggs to hatch early, the lambs may still be too young to be grazing, and could escape infection.

Teladorsagia circumcincta (causes ostertagiasis/ostertagiosis)

A small brownish roundworm which depresses appetite by invading the gastric glands of the stomach. This parasite is responsible for causing diarrhoea and general illness, slowing growth rates, and can kill weakened lambs.

Haemonchus contortus (causes haemonchosis)

This is one of the most pathogenic nematodes. Also known as the 'barber's pole worm' because of its distinctive striping, this worm causes anaemia in sheep because it attaches itself to the lining of the abomasum and ingests large quantities of the host's blood. Sheep of all ages are vulnerable and the worms lay vast quantities of eggs. Warmer weather is required for the larvae to develop, so haemonchosis is often more apparent in late summer and early autumn.

Trichostrongylus spp. (cause trichostrongylosis)

This is sometimes called the 'hair worm' because of its strand-like appearance, or the 'black scour worm' because it causes diarrhoea. This intestinal worm often hits store lambs in the autumn months, but can affect them earlier. As its common name - the 'black scour worm' – suggests, it causes diarrhoea, it inflames the intestines, results in rapid weight loss, and death.

Tapeworm

Tapeworms are flat, segmented, parasitic worms and it's not unusual to see segments in sheep faeces. Sheep are one of the host species for tapeworm, which attach to the intestines. Adults rarely appear to suffer ill-effects, but in lambs, an infestation can be fatal. Infection can start when grazing lambs consume infected pasture mites, which are also hosts. The larvae then mature in the intestines. Weight loss can occur and particularly vulnerable lambs can die. Most modern wormers will treat both roundworm and tapeworm.

Dogs also act as tapeworm hosts, so there is concern that sheepdogs eating sheep faeces will become infected and continue the life cycle. Increased awareness of the need to worm dogs regularly is, however, reducing the problem.

Lungworm

Although not considered as great a threat as other parasites, lungworm can cause parasitic bronchitis, resulting in coughing and loss of condition. Some sheep will show no signs of infection at all. Sheep with Johne's disease may have heavy lungworm burden, as a result of the disease having an impact on the immune system. As with roundworm, many of the wormers on the market will also kill lungworm, as will some of the fluke treatments.

LIVER FLUKE DISEASE (FASCIOLASIS)

This disease of sheep and cattle is caused by a small, flat parasite – *Fasciola hepatica* – but the life cycle is very different to that of other parasites which use ruminants as a host, because the mammal affected by the illness is not the only host involved in the cycle. The other organism is a small snail – *Limnea truncatula* – which plays an equally important role. The snail loves muddy, slightly acidic ground, so areas that are poorly drained, or which have heavy rainfall, are more prone to problems.

The life cycle doesn't really have a beginning because it's a continuous process! However, let's start, for argument's sake, with the sheep.

Eggs from adult female fluke are shed in sheep faeces and contaminate pastures. When the ground is damp and the temperature between 7°C and 10°C, the eggs hatch and the larvae go looking for the snails, which will be their hosts. Once inside the snail, the larvae develop and eventually leave the snail, swimming until they reach vegetation. They undergo a further transformation, developing into dormant cysts – metacercariae. When the sheep eats the vegetation, immature fluke are released from the cyst and penetrate the intestinal wall, eventually making their way to the liver. It takes four or five weeks for the immature fluke to migrate through liver tissues before settling in bile ducts. In all, the time taken between getting into the sheep's body and making an impact on it is eight to 12 weeks. Immature fluke burrow through the liver causing serious damage, while adults reside in the bile ducts where they ingest blood, causing anaemia.

Sheep that previously looked healthy can be found dead without any symptoms being noticed, the cause of death being haemorrhage and liver damage. Close inspection of the flock may reveal others looking lethargic and being reluctant to graze. August to October is normally the key period, though climate change and geographical variations mean shepherds should stay on the alert outside this window.

Treatment is normally by drenching with triclabendazole and the flock should then be moved to clean pasture or re-treated every three weeks for the next three months. Your vet will advise the best course of action. In all cases, good nutrition will help recovery.

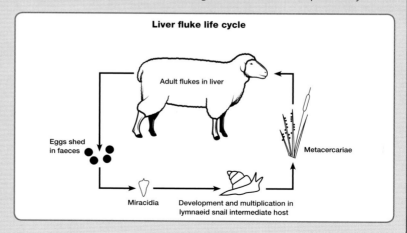

Liver fluke life cycle

Adult flukes in liver

Eggs shed in faeces

Metacercariae

Miracidia

Development and multiplication in lymnaeid snail intermediate host

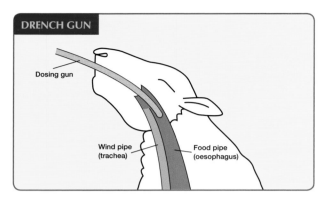

DRENCH GUN

Dosing gun

Wind pipe (trachea)

Food pipe (oesophagus)

Administering wormers

If you're unsure about how to administer any drug, ask your vet to show you. Failing to give treatments properly is not only a waste of your time and effort but puts your sheep through unnecessary stress and also means you are throwing money down the drain.

Wormers come in two types: those that are injected and those that are given as an oral 'drench'. Injectable wormers are given subcutaneously (under the skin) – see oage 59 – while oral preparations are loaded into a drench gun, which is carefully positioned in the mouth of the sheep.

GIVING AN ORAL WORMER
- Weigh the biggest sheep in your flock and calculate the dose you need to give based on that weight, so that you don't risk under-dosing.
- Make sure your drench gun is working accurately and delivering the correct dose before you start. Always wash and dry the equipment thoroughly after use.
- Use the gun correctly. Sheep need to be restrained properly to prevent injury, and to make sure they swallow the wormer. Hold the head and tilt it slightly to one side, insert the long nozzle between the molars and the incisors and over the back of the tongue. Simply squirting the product into the mouth isn't good enough, because it will escape down the oesophageal groove and so by-pass the rumen.

Other ways of combatting worms

Anthelmintics are just part of the way in which worm burdens can be reduced and experts advise that good management of pasture and stock, planned rotation of stock, along with appropriate nutrition, should all accompany any drug treatments:
- **Grazing lambs**. Lambs can be weaned from 12 weeks and moved to pastures known to be less heavily contaminated by worms. It's also worth considering dividing lambs into age groups to minimise the risk of older lambs shedding eggs and infecting younger, less-resilient ones.
- **Using ewes for cleaning up land**. As mature ewes in good condition generally have low worm burdens, they can be used to reduce contamination on high-risk pastures by ingesting infected larvae.

- **Cross-grazing**. Grazing cattle alongside sheep – or allowing them onto pasture once the ewes and lambs have left – can help reduce worms as they ingest the larvae but aren't affected by them, as the parasites are host-specific.
- **Nutrition**. Sheep with nutritional stress are more vulnerable to the effect of internal parasites, so maintaining good condition is important. Research suggests that ewes fed fodder containing high levels of nondegradable protein shed fewer worm eggs in their dung. Giving lambs creep feed not only helps give them a good start in life but also helps them to be more resistant and means they are less likely to graze and ingest large quantities of larvae.
- **Grazing on alternative crops**. Some grazing experiments have shown that grazing on bioactive forages – such as birdsfoot trefoil and chicory – may reduce the impact of internal worms on sheep.

External parasites

As with internal parasites – and a whole host of diseases – external parasites can be brought in when you buy new stock, so quarantine new arrivals for up to four weeks and treat them for sheep scab and internal parasites at the same time. Any signs of scratching, rubbing, loss of wool, or twisting round to attempt to chew the fleece should set the alarm bells ringing and prompt you to examine the sheep concerned.

Sheep scab

Sheep scab is a type of mange caused largely by a burrowing mite, *Psoroptes ovis,* which attacks the woolled areas. Another mite, *Sarcoptes scabei var. ovis*, causes sarcoptic mange and concentrates on the areas without wool, such as the head, neck and legs, but is rarely seen in the UK. Until compulsory dipping of sheep was banned in 1989, sheep scab was rarely seen, and there is now growing opinion that it should be reintroduced.

Below: Wool loss due to sheep scab.

Above: *Ixodes ricinus.*

The mites cause skin lesions, which develop into pustules that rupture and discharge a sticky liquid, which causes crusty scabs. Sheep scab is considered a winter problem, spanning the period October to March, but cases have been identified at other times of the year.

Correct diagnosis is important, as scratching and rubbing can also be caused by lice and other infections. Once the cause has been established, sheep should be dipped or injected according to veterinary advice, and all sheep should be treated at the same time.

It's worth knowing that sheep scab mites can survive off the sheep for up to 17 days, and they last longer if the weather is cold and damp. Bear in mind that straw and hay can, therefore, harbour mites – as can buildings and trailers.

This remains a notifiable disease in Scotland.

Lice

As with scab, lice normally begin to cause problems in winter and, like sheep scab mites, can live off the sheep for more than two weeks, surviving on wool and dead skin cells. Biosecurity measures like those suggested for scab should be applied if new stock is being bought in. Identifying whether parasites are scab mites or lice can be tricky in the early stages, but lice don't cause the sheep to lose wool in such dramatically large patches as do mites.

Unlike scab, lice problems can be treated using pour-on preparations containing deltamethrin, cypermethrin or alphacypermethrin. Dipping, under veterinary supervision, may also be an option.

Ticks

Ticks – particularly the sheep tick, *Ixodes ricinus* – prefer damp habitats with deep vegetation, such as areas of rough grazing, heath, woodland and moorland, and the uplands of the UK have seen a significant increase in recent years.

This parasite can carry:

- Tick pyaemia – this is sometimes known as 'cripples' and affects lambs from 2 to 12 weeks old, causing crippling lameness and paralysis.
- Tick-borne fever – a bacterial disease that can affect fertility, cause abortion and lower resistance to respiratory and joint infections.

- Louping ill – a viral disease that affects the brain, causing lack of coordination, convulsions and sudden death.
- Lyme disease – a bacterial disease causing pain and swelling in the joints, stiffness and lameness.

The tick lays thousands of eggs and while the larvae and nymphs feed on young lambs and other small mammals the adults concentrate on larger mammals – including humans. The risk period is quite long, with key times for activity being February to October but peaking mid-summer if conditions are humid.

Irritation caused by ticks can be confused with sheep scab or lice, so correct diagnosis is important. They tend to attach to the face, ears, the inside of the leg and the groin and begin sucking blood. If left undisturbed, they can feed for up to seven days. Protection against ticks can be achieved by dipping and pour-on chemical treatments recommended by your vet.

Keds

Melophagus ovinus is a wingless fly that lives on sheep and sucks blood, causing anaemia and loss of condition. A substance it excretes can also stain the fleece, reducing the value. Keds can be distinguished from ticks by the number of legs – they have six, whereas ticks have eight. Effective control is achieved by dipping, but pour-on treatments are also available.

Flies and flystrike

For all sheep farmers, large or small, this is the most likely form of ectoparasite problem you will encounter.

Flystrike, blowfly strike, or simply strike (myiasis) is caused by maggots of greenbottle, bluebottle or black blowflies infecting the flesh and, if not noticed sufficiently early, sheep can die within days.

The flies are particularly attracted to open wounds and to thick fleeces contaminated with faeces or urine; they lay eggs that hatch within 12 hours and the resulting larvae begin

Below: The band of missing hair on this ewe is due to flystrike.

Above: Maggots in a fleece.

Jan Walton

eating their way into the flesh, causing severe irritation and toxaemia – blood poisoning as a result of bacterial toxins.

One of the prime targets is the back end of the sheep, where faeces are most likely to collect, though the sides of the sheep are often attacked because, when the sheep lies down, the hind feet, which may be covered in faeces, rub

Below: Preventative treatments.

against the fleece. The head and feet can also be affected, and sheep with foot lesions should be checked regularly for evidence of infestation.

Symptoms to watch for

Signs of strike can be difficult to spot without detailed examination, but affected sheep may break away from the rest of the flock and seek out quiet areas in long vegetation. The irritation will cause them to rub against fences, gates and trees, and you may see them turning around to try and bite the affected area. They may be seen stamping their legs, flicking their tails, and will be increasingly restless.

In light-coloured sheep, you may notice a darker patch where the sheep has been struck; on examination, the area will feel damp, as the larvae exude enzymes to help digest the flesh. On parting the fleece, maggots will almost certainly be visible and there will be a foul smell. Some flies that would not initiate a strike are, however, attracted to the open wound and exacerbate the problem.

Strike can start as soon as the warm, humid weather arrives – sometimes as early as March or April – and can last until as late as December. Upland areas may see a shorter strike season.

Preventing strike

Protecting against strike makes far more sense than trying to eradicate the problem once it's taken hold. Good husbandry is essential, so keep a close eye on your sheep during the fly season, inspecting twice a day if possible. Shearing (see Chapter 7) significantly reduces the risk of strike, but if sheep can't be sheared early in the season then crutching or dagging – removing wool from around the anus and tail – should definitely be carried out before the flies arrive. Be aware that even sheep that have been sheared can still be susceptible to strike, particularly if shearing takes place early in the year and the fleece starts to grow back. Lambs, too, can be hit, with long-woolled breeds especially vulnerable.

Flies are more likely to be attracted to sheep that are scouring, possibly due to internal parasites, so make sure they don't have a worm burden by getting a faecal egg count done, and treat if necessary.

Pour-on insecticides come in two types – those that just act as a deterrent to flies and those that will not only repel flies but will also kill existing maggots. If you need to kill maggots fast, choose the right product for the job. However, you need to be aware that products that kill maggots will generally repel flies for a much shorter time than those products designed solely as deterrents. Make sure you read the instructions thoroughly and know roughly how much protection the products will offer – but be aware that heavy rain can have an impact on their effectiveness. Damp sheep can be treated, but don't apply in heavy rain or if heavy rainfall is expected.

Pour-ons should ideally be applied using a compatible applicator from the manufacturer, which has a fan-shaped head attached for wider coverage.

Above: Start applying the pour-on at the neck and work down to the rump.

Below: After applying down the spine, continue spraying down the hind legs.

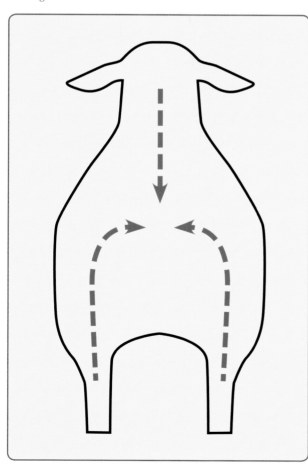

Dealing with strike

Swift action is required when a case of flystrike is spotted. Chances are, if one sheep is affected, others in the flock will be, too.

1 Wearing suitable protective clothing, shear or clip affected areas to see what you are dealing with. Trim back well, as close to the skin as possible, and, once the affected area is identified, take the wool back a further 5cm all round.
2 Clean out any maggots you can see and carefully place any wool containing maggots into a plastic bag and leave for a few days until they die.
3 Apply a suitable flystrike treatment to the clipped area. As well as seeing surface maggots squirming, you will also see many more wriggling their way out of the skin. This may sound gruesome, but it's quite satisfying in a weird way, because you know at this point that you're doing something that will not only relieve the sheep's discomfort but will potentially save its life. Pour-ons are most popular, but summer fly cream (see page 78 in Chapter 7) and maggot oil can also be used.
4 Give a long-acting antibiotic injection to fight potential bacterial infection.
5 Move any treated sheep into a separate paddock or building so that they can recover and be monitored. This also reduces the risk of others being infected, as areas struck by maggots will attract more flies. Areas treated will lose wool, but it will start regrowing in time.
6 Make sure you note down any withdrawal times specified – either for meat or milk. Some newer treatments have a zero-day withdrawal period, but check to make sure. Log details of all treatments in your veterinary records.
7 When everything has been treated and you've washed and dried your equipment, consider what to do in future regarding affected sheep. The latest advice is that if particular sheep are repeatedly being hit by flystrike when others are not then you should consider culling. The thinking is that resistance (and susceptibility) may be inherited.

DISPOSING OF DEAD STOCK

If any animals die on your holding – or are stillborn – you must dispose of them legally. It's against the law to bury or burn carcasses on site, so you must either arrange collection by an approved collection company or a local hunt with its own incinerator. All of this will normally be at a cost. While you are waiting for collection, you must protect the carcass from being eaten by birds and other predators. Failure to dispose of a carcass properly can result in prosecution.

SHEARING

Depending on where you live – and often on the availability of the shearer – shearing is carried out from mid-May to mid-summer. It's normal to aim to shear before the flies begin to lay eggs, but sometimes this has to be delayed and preventative treatments need to be used as a temporary measure. Pregnant ewes are not normally sheared before lambing, to avoid unnecessary stress, though some may be in early pregnancy if they are to be housed in close conditions. Certain breeds that are being raised specifically for their highly prized fleeces are left until later in the year – sometimes until early winter – in order to add value to the crop.

If you visit some of the summer agricultural shows, you may see unshorn sheep because specific breed classes insist they must be shown 'in the wool'. Another factor in deciding when to shear is knowing the time when the fibres in the older part of the fleece naturally begin to loosen or 'rise' and separate from the new growth. Attempting to shear a sheep whose wool has not 'risen' can be tricky and can damage equipment. Similar resistance is encountered when trying to shear lambs.

Above: Hillbreeds like the Herdwick (left) and the Badger Face produce wool with interesting colour variations.

Man has been shearing sheep and using their wool for thousands of years. Shearing provides important welfare benefits, helping to make the sheep more comfortable in hot weather and minimise the risk of overheating, but particularly to reduce the chance of flystrike.

'NO-SHEAR' SHEEP

If you're not particularly bothered about getting money for your wool and would like to minimise shepherding tasks, there are sheep that shed their wool naturally and so don't need to be sheared. Breeds including the Wiltshire Horn, the Dorper and the Barbados Blackbelly, plus newer hybrids such as the Easy Care and the Exlana, are all sheep that start to lose their coats once the milder weather arrives.

Below: Wiltshire Horn ewes starting to shed their coats.

British Wool Marketing Board

The British Wool Marketing Board exists to help producers get the best price for their fleeces. Wool prices fluctuate throughout the year, but the average price at the time of writing (2015) was around £1.50 per kg. The board has its headquarters in Bradford, West Yorkshire, but has offices in England, Scotland, Wales and Northern Ireland. Contact details for the regions are on their website www.britishwool.org.uk or can be obtained by ringing 01274 688666.

Despite wool being such a valuable commodity in our ancestors' days, today's prices are nothing for the average small-scale sheep keeper to write home about. If you have a flock of hundreds – or even thousands – of sheep, and can shear them yourself, you may well turn a small profit. However, for the smallholder with just a handful or even a few dozen, the cost of hiring in a shearer can cancel out any money received for fleeces.

The British Wool Marketing Scheme of 1950 requires all those with more than four sheep to register with the British Wool Marketing Board (BWMB). Sheep owners in the Shetland Islands have an exemption, as they have a separate arrangement for selling fleeces. The board isn't a profit-making organisation and markets the wool of producers by selling at auction. On registration, the board will put you in touch with a local depot, which will supply wool bags and arrange collection. Fleeces fall into different price categories according to breeds and will also be graded according to their quality. If you want to keep your own fleeces and process them yourself, you can apply for an exemption.

Above: Setting up a professional shearing system.

Shearing contractor or DIY?

Good shearers are often booked up far in advance and it can be difficult to find one who is willing to shear just a handful of sheep. If you can find a few other small-scale shepherds in your area, it might prove a more attractive prospect – but organising how, where and when they are penned, and for how long (bearing in mind standstill regulations) might prove problematic.

You might decide that you want to learn to shear yourself, and there are plenty of courses available, including some run by the BWMB. If you have a few friends with small flocks who also fancy shearing their own, you might want to get some training done and then arrange to help one another on different days.

When you see professional shearers turn up at farms or taking part in shearing competitions, they will inevitably be using some expensive kit with overhead machine systems and heavy-duty handpieces. As with most professions, if you're serious about your work, you need to invest a significant amount in the tools of your trade, and with shearing equipment, you could be talking several thousand pounds for the perfect set-up.

Inexpensive electric hand-shearers are available on the Internet for less than £100, but don't expect them to last a lifetime. They should be fine for doing a dozen or so sheep now and again, but they aren't built to shear hundreds of sheep in a short space of time; overuse can quickly burn out a cheap motor. Some shearing kits work off mains power, but there are an increasing number of good lightweight 12-volt shearing kits on the market, which can work off a car battery or generator, allowing you to shear anywhere you please.

Occasionally, shearing kits come up for sale at auction, but you need to understand what you are buying and you can never be sure of how well they have been maintained. You might want to consider buying equipment jointly with others in your area and sharing the benefits – and maintenance costs. When buying shearing equipment, remember that the handpiece is just one part, and you will also need cutters and comb to suit the type of fleece you are dealing with. Good suppliers will be happy to help with advice but also talk to other sheep owners with the same breed to find out what works for them.

Hand-clipping

Just because you see burly men in vests at agricultural shows shearing sheep after sheep in a matter of seconds doesn't mean that there is a minimum speed requirement. Provided the sheep is comfortable and not stressed, you can clip slowly and steadily using hand shears. The end result – and the quality of the fleece – will probably not be as neat and tidy as with electric-powered clippers, but the job will be done. If you're not worried about how your fleece will look, but just want to make your sheep more comfortable, you don't need to follow any particular pattern of clipping – just work the way that suits you best. Traditional shears are often the first choice for hand-clipping, but newer, super-sharp, scissor-type shears will make clipping a lot easier.

Shearing – how the experts do it

For obvious reasons, it really is important to be taught how to shear safely and properly, so don't skimp on tuition. The sequence of pictures on the following pages is meant as an overview of the process, rather than a 'how to' guide. There is no substitute for proper training.

Just with other rural skills, such as hedge-laying and drystone walling, there are different styles of shearing and shearers also develop their own particular ways of doing things, so methods will vary.

Prepare your shearing area, using a barn or an area with shade from sunlight. A large piece of hardboard might be useful as a surface on which to stand. The same basic principles apply, whether you are using electrical shearing equipment or simple hand-clippers. Ensure that all cutting

equipment is sharp and well oiled. Keep some oil and a few cloths on hand, just in case. Mechanical equipment should be well maintained and working correctly. Make sure you have a tarpaulin, a table, or a clean piece of board ready for laying out and rolling the fleece. Get everything set up before you bring in your sheep. Don't keep them hanging around getting stressed.

As with any job involving livestock or potentially dangerous equipment, suitable clothing is important. Professional shearers wear traditional suede moccasins (below) because

the amount of lanolin in the fleece can make work surfaces extremely slippery for normal shoes or boots; natural leather grips the floor more easily. They also wear purpose-made trousers, which are made of a kind of double-strength denim. Why? Well, if you're handling horned sheep, you run the risk of bruised and battered legs. Also, when you're dealing with a large number of sheep, day in, day out, contact with lanolin can cause skin irritation.

Assemble a collection of veterinary/husbandry items you might need, should any problem (e.g. foot rot, flystrike, lesions, mastitis) be spotted during handling. Your kit should include things like an antibacterial spray, an injectable antibiotic, needles and syringes, and foot-trimmers. Super-fast, experienced shearers might also include a suture kit for closing bad wounds caused during shearing, but, if you're taking the cautious approach, the odd nick will be treated easily with a quick spray.

Be prepared to find maggots – even if you have used preventative pour-ons, flystrike can still occur – and have suitable treatments ready just in case. This is also a good time to treat lambs with a pour-on and trim messy back ends, which can attract flies. You may also want to carry out routine husbandry tasks such as worming or vaccinating while you have your sheep rounded up. Keep a notebook close to hand to log anything unusual, and to record any veterinary treatments given. The details should then be transferred to your farm veterinary records.

Round your sheep up into a holding pen or race – maybe a temporary enclosure made from hurdles – having first checked that their fleeces are dry. A dry fleece is important for a few reasons: firstly, if you're using electrical shearers, you might electrocute yourself; it makes shearing more difficult; and a wet fleece can deteriorate when stored. If you're hiring in someone to do the job, don't waste their time. Have the flock penned up and ready to go.

With the sheep safely restrained, the shearer starts by trimming off the wool on the brisket and the belly, taking care to protect the teats and genitals with his spare hand. The wool on the brisket and belly tends to be of very poor quality and is normally discarded. He continues downwards, taking off any wool on the inside of the legs and the tail.

The left flank is next, with the shearer working up the leg with long strokes, right to the spine or to the other side of it if possible. He uses his spare hand to press down into the stifle joint in order to straighten out the leg, pull the skin tighter, and make shearing smoother and safer. A long stroke is made up the spine, from the tail to as far as the shearer can reach.

The sheep is turned into a more upright position to allow the throat and the neck to be sheared. The shearer starts at the brisket and works upwards towards the cheekbones, around the left ear, and up the neck.

Next is the left foreleg, with the shearer straightening out the leg and working back up towards the shoulder. The sheep is rotated and laid down so the shearer can use a series of long strokes up towards the neck. This can be a tricky step, as the sheep must be kept under control and not allowed to struggle to its feet. Rolling the sheep slightly towards him, the

shearer makes long strokes along the sheep's back, with the aim of meeting the point at the spine where the other side of the fleece has already been removed.

After completing the strokes along the back, the sheep is repositioned slightly again to allow any wool on the right side of the head and around the ear to be removed. The wool on the right shoulder and down the right foreleg and the rest of the right-hand side of the fleece is then removed, working down the flank and finishing on the lower part of the rear leg.

Rolling your fleece

Rolling a fleece is not only a way of keeping it neat and tidy and making it easier for storage – it's also the way that wool buyers will expect to see it presented. Preparing a fleece correctly is a real art, and there are local, national and even international competitions to find the fastest and neatest rollers.

As mentioned earlier, you will need a clean surface on which to work. Shake the fleece out flat, so that the skin side is down and the outer side of the fleece is facing you. The only exceptions to this rule are fleeces from Blackface, Herdwick and Rough Fell breeds, which should be laid skin side up.

The neck end of the fleece should always be furthest away from you. Remove any bits of dirty wool, straw and other foreign bodies. If the small bits of belly wool are clean enough, place them in the centre of the fleece. If not, dispose of them. Turn the flanks, followed by the britch (tail) end of the fleece inwards and begin rolling towards the opposite end, aiming to keep everything tight and even. When you reach the narrower neck wool, you can either tuck the end inside the bale (the method preferred by the British Wool Marketing Board), or twist it to form a rope shape and wrap around the fleece to secure it.

Maximise the value of your fleece by caring for it when it's still on the live sheep. Colouring fleeces with bloom for showing (see page 147) will make it worth less, as will overuse of stock-marking spray. Similarly, coloured cord or twine (e.g. used to secure a prolapse harness) can leach unwanted pigment.

FEEDING AND NUTRITION

How ruminants work

Whatever animals you keep, you need to understand their nutritional needs. Correct feeding is essential for everyday maintenance of condition, as well as for successful growth and reproduction. Nutritional needs vary at key stages of life and the ability to assess what is required will be made a lot easier with a little understanding of the way in which anything eaten is processed.

Sheep, like cows, goats, llamas and alpacas – and more exotic species like water buffalo, bison, giraffes, deer, antelope and camels – are ruminants. They are herbivores equipped with a fascinating four-chambered stomach, the four chambers being the rumen, the reticulum, the omasum and the abomasum.

The ruminant's complex and extremely specialist digestive system has evolved in order to be able to break down a fibrous diet (e.g. tough vegetation), which would otherwise be difficult to process. In the wild, a ruminant would survive solely on raw, fibrous plant matter, so it has become specially adapted to get the best out of it.

As with all mammals, the digestive process begins in the mouth. The first part of the process is mechanical: food is cut and torn in the mouth and mixed with saliva. As mentioned in Chapter 4, sheep only have sharp cutting teeth in the bottom jaw. In the top jaw, there are flat molars and a hard pad, so the jaw makes a rotary movement to grind the food rather than crush or tear it. The molars are shaped and spaced in a way that allows the animal to chew on one side of its jaw at a time, which helps in shredding tough plant fibres. The salivary glands are important because they produce vast amounts of saliva, needed to fully permeate the food during rumination ('chewing the cud').

During grazing, plant matter is passed into the first chamber of the stomach – the rumen, a large 'fermentation vat'. The sheep regurgitates small amounts of partly chewed

DIGESTION IN LAMBS

In young ruminants, the rumen and the reticulum are not fully developed; when milk reaches the stomach, it is sent via a tubular fold of tissue (the oesophageal or reticular groove) directly to the third and fourth compartments – the omasum and abomasum. As the lamb (or other ruminant) starts to eat solid food, the first two compartments grow in size. In the adult, these two compartments comprise 85% of the total capacity of the stomach. The oesophageal groove will not normally function in the mature animal, though the groove can still be stimulated to close to form a channel if, for instance, a bottle feed is given.

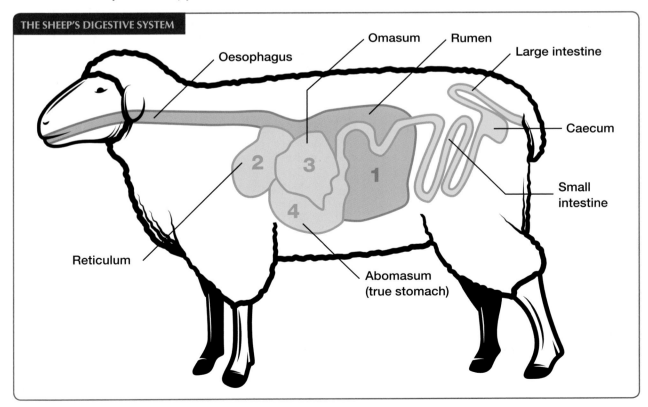

THE SHEEP'S DIGESTIVE SYSTEM

Oesophagus

Omasum

Rumen

Large intestine

Caecum

Small intestine

Reticulum

2

3

1

4

Abomasum (true stomach)

Right: Assessing condition correctly is a skill that comes with practice.

food (the cud) back into the mouth, where it is chewed and mixed with more saliva. Every piece of food is regurgitated and chewed up to 50 times, and the time spent ruminating depends on the fibre content. Sheep saliva doesn't contain amylase, the enzyme in some mammals that helps turn starch into sugars, so the purpose of chewing the cud is thought to be to aerate, macerate and mix the food with saliva more thoroughly to aid digestion.

Only when solids are small enough (less than 1mm) will they be passed into the reticulum – a much smaller chamber. Once the feed has been broken down into sufficiently small particles, it passes into the omasum, where water and nutrients are absorbed.

The fourth and final chamber is the abomasum, known as the 'true' stomach. The abomasum produces hydrochloric acid and digestive enzymes, such as pepsin, which breaks down proteins.

The next stages of the digestive process take place in the small and large intestines, where further nutrient absorption takes place. Any remaining matter is excreted.

Understanding feeding requirements

How much feed does a sheep need? Well, it's one of those 'piece of string' questions: it depends! If there were a simple guide to feeding sheep, life would be a lot easier. Unfortunately, there isn't one – but then there isn't a foolproof blueprint for humans, either. What you're aiming to do at all times is to provide your sheep with sufficient

food that they stay in good condition, but without risking them putting on too much weight. Different breeds of sheep can vary incredibly in terms of nutritional needs: whether they are classified as a hill, upland, or lowland breed will have a bearing; different environmental challenges can have an effect; and, of course, as mentioned earlier, key stages in life put additional demands on the body.

Sheep that have access to sufficient good-quality forage (be it grass, hay, silage, haylage, or root crops) all year round should have no problem keeping their weight on, but you need to know how to check whether your sheep are getting enough of what they need.

The tried and tested method of assessing this is by 'condition scoring'. This is – literally – a 'hands-on' method of working out whether your sheep are too thin, too fat, or just right. What you are doing is feeling for the amount of covering over the top of the vertebrae (spinous process), the small bony projections on the right and left side of the vertebrae (transverse processes) and the bones in the tail.

Once you've familiarised yourself with the technique and determined how your sheep are doing, you can increase or reduce their feed as appropriate.

There are three key positions on the sheep where condition scoring is carried out. These are the dock of the tail, the loin and the final two 'short ribs' – often known as the 'ten to two' position, as in the hands of a clock (see illustration overleaf). In addition, the breast can be felt, as can the top inside part of the hind legs.

There is an internationally accepted way of rating condition using a scale of 1 to 5 – with 1 being emaciated, 3 being average condition, and 5 being obese. You will come across experienced shepherds using half-scores (0.5), but don't get too worried about differentiating to that level of detail when you're just starting off.

A) THE DOCK OF THE TAIL

- Condition Score 1: Individual bones of the tail felt very easily.
- Condition Score 2: Bones easy to feel with light pressure.
- Condition Score 3: Bones felt with moderate pressure.
- Condition Score 4: Firm pressure needed.
- Condition Score 5: Bones cannot be felt.

B) THE LOIN

- Condition Score 1: Transverse processes very prominent; easy to feel between the ribs and the squarish ends of the ribs.
- Condition Score 2: Spinous and transverse processes very prominent; ends of the ribs are more rounded.
- Condition Score 3: Tips of processes feel rounded and individual bones can be felt with light pressure.
- Condition Score 4: Spinous processes can be felt with moderate pressure; transverse processes felt with firm pressure.
- Condition Score 5: Individual processes cannot be detected.

C) 'TEN TO TWO' POSITION (FINAL TWO RIBS)

- Condition Score 1: Individual ribs prominent and easy to feel.
- Condition Score 2: Slight covering of flesh over ribs, but still easy to feel.
- Condition Score 3: More fat cover over and between ribs; less easy to feel.
- Condition Score 4: Firm pressure needed to detect any individual ribs.
- Condition Score 5: Too much fat to detect individual ribs.

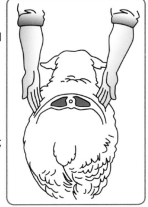

CONDITION SCORE TARGETS

	Hill ewes	Upland ewes	Lowland ewes	Rams
At weaning	2	2	2.5	
At mating	2.5	3	3.5	3.5
At lambing	2	2.5	3	

Condition Score 1
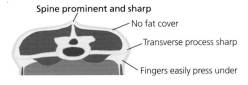
Spine prominent and sharp
- No fat cover
- Transverse process sharp
- Fingers easily press under

Condition Score 2
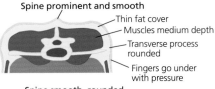
Spine prominent and smooth
- Thin fat cover
- Muscles medium depth
- Transverse process rounded
- Fingers go under with pressure

Condition Score 3
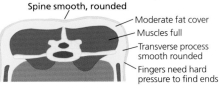
Spine smooth, rounded
- Moderate fat cover
- Muscles full
- Transverse process smooth rounded
- Fingers need hard pressure to find ends

Condition Score 4
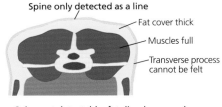
Spine only detected as a line
- Fat cover thick
- Muscles full
- Transverse process cannot be felt

Condition Score 5
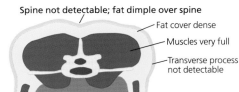
Spine not detectable; fat dimple over spine
- Fat cover dense
- Muscles very full
- Transverse process not detectable

THE IMPORTANCE OF WATER

Never underestimate the part water plays in maintaining health. Water is the main constituent in the body of any mammal, comprising as much as 50 to 80% of its live weight. Although significant amounts of body weight can be lost without the animal dying, a loss of 10% of the normal level of water in the body can be fatal. Depending on the moisture content of what it is eating, a sheep will need at least 2.5 litres a day – more if being fed concentrates or hay. Lactating ewes can need as much as 8 litres a day. Sheep don't like drinking contaminated water, so keep drinking troughs clean. They also prefer moving water (e.g. a stream or a running hosepipe) to anything provided in containers.

Supplementary feeding

Provided you have not overstocked your fields, and provided you have maintained your grassland well (see Chapter 3), your sheep shouldn't need much – if any – supplementary feeding through the milder months, when grass growth should be at its best. However, if there is insufficient grass, you will have to offer some form of additional feed. Hay is the most popular preserved forage, but silage and haylage (see Chapter 3) are also used.

There are risks to feeding silage and haylage, as both can harbour bacteria and fungi. The biggest concern, particularly with silage, is the risk of listeria – a type of bacteria found in soil, which will multiply inside a silage bale where it has nutrients, moisture and warmth to multiply. Development is slowed down if the silage has been efficiently fermented, a low pH maintained, and if the amount of oxygen inside the bale has been minimised by careful, multi-layered wrapping. However, if, for instance, a bale is punctured – maybe while being moved, or if it has been damaged by birds or rodents – oxygen and moisture will creep in, helping bacterial growth. Listeria can cause a variety of conditions, including neurological problems and abortion.

Common-sense tips for feeding silage

- If you are making it yourself, aim to remove as much air as possible when wrapping and make sure the wrap is secure, using several layers.
- Store bales safely and try to minimise the risk of attack by birds and vermin.
- Take care not to damage silage wrap when transporting.
- Throw away any silage that looks damp, mouldy, or contaminated with soil.
- Remove any leftover silage every two days and keep feeders clean.

Root crops as forage

As mentioned in Chapter 3, root crops can be a valuable source of fodder and they are quick and cheap to grow, too. Most crops will only take two to three months before they can be grazed or harvested, and they can be an extremely useful alternative to bought-in feed when grass supply is short.

Some roots, such as stubble turnips and swede, can be grazed where they grow, but others, like fodder beet, need to be lifted and cleaned before feeding. Although most sheep will enjoy eating the entire plant, harder roots like swede should not be fed to sheep at either end of the age spectrum – as immature teeth in lambs and weak or wobbly teeth in older ewes can easily be broken.

As with any new food, roots should be introduced gradually, to allow the rumen to adapt. Make sure, too, that if you are allowing sheep to graze on a growing crop, you restrict access using electric fencing to avoid too much trampling, as sheep don't like to eat soiled food.

Potential adverse side effects

There are some issues of which to be aware when feeding root crops, particularly those of the brassica family. Many farmers feed their sheep year in, year out without any problems, but there are potential risks in certain situations.

- **Photo-sensitisation**. Sheep with white heads, ears, or faces can become hypersensitive to sunlight, with burning, blistering and scabs. This sometimes occurs when crops are grazed too early, and it is most often seen with rape and kale.
- **Nitrate poisoning.** Crops grown in soil containing high levels of nitrates can experience a build-up of the compound in the leaves. Sheep can suffer symptoms including abdominal pain, scouring, weakness, muscle tremors and drooling. In severe cases, they can become comatose and die.
- **Goitre.** This swelling of the thyroid glands is caused by a lack of iodine. Brassicas contain glucosinolates, which block the uptake of iodine by the thyroid. Iodine deficiency can affect fertility and cause stillbirths and death in newborn lambs. If a deficiency is diagnosed, iodine supplements can be given, either by injection or in a mineral block or bucket.
- **Kale anaemia (redwater).** Signs include weakness, red urine, reduced appetite, loss of condition and poor fertility. Fertilisers containing sulphur should be avoided and sheep should have adequate copper and selenium levels in the diet.
- **Bloat.** Brassicas are rapidly broken down in the rumen and can cause this condition in which the stomach fills with gas and swells because the sheep can't release it by normal means – i.e. belching. Risks can be reduced by providing fibre alongside the root crop and allowing only gradual access to avoid overconsumption.

FEEDING TIME

Give your flock sufficient space to avoid some getting pushed out at feeding time. Don't forget that horned sheep will need even more space. If you can, feed twice a day – particularly in the run-up to lambing.

Feeding concentrates

If hay or silage is in short supply, a concentrated – or 'compound' – feed is usually the first choice as a supplement. Forage must remain the main part of the diet, otherwise sheep can be prone to acidosis, which, as the name suggests, is an indication of high levels of acidity. The rumen needs to keep a pH of 6.2 to 6.5 in order for the microflora to function properly. If high levels of concentrates or straight cereals are fed, lactic acid causes the pH in the rumen to drop significantly. The low (acidic) pH leads to a decline in microflora, which also become less active and less efficient. Less fibre is digested and appetite is depressed, risking loss of condition and reduced milk production in lactating ewes.

When introducing concentrated feed or cereals, be careful to do so gradually, to avoid upsetting the rumen. There are numerous types of commercial compound feeds available, formulated to suit various stages of the sheep's life, including lamb 'creep' pellets, weaning diets, grower feeds, high-protein feeds for pre- and post-lambing, as well as coarse mixes (which look a bit like muesli). Some manufacturers will offer more than a dozen different products, so it can be difficult to choose the right feed for the right time. Most bagged feed will come with instructions on how much and how often to feed. You will often find more information on the manufacturer's website, including a breakdown of the formulation. All of the big manufacturers employ specialist advisers who will be happy to point you towards the most suitable product. Your local feed merchants or agricultural store may also be able to offer advice.

Essential minerals

There are more than a dozen minerals that are considered essential for good health in sheep. The main ones are sodium (i.e. salt – chemical symbol Na), chloride (Cl), calcium (Ca), phosphorus (P), magnesium (Mg), potassium (K) and sulphur (S). Some are required in small amounts, including iodine (I), copper (Cu), iron (Fe), manganese (Mn), zinc (Zn), molybdenum (Mo), cobalt (Co), selenium (Se) and fluoride (Fl).

Insufficient amounts of sodium in the body can result in lower consumption of feed and water, a drop in milk production and poor growth of lambs. Signs of sodium deficiency can be sheep chewing wood or licking the earth. Calcium and phosphorus deficiencies may lead to skeletal problems, including rickets, while an imbalance of these two in the diet can cause urinary calculi (see panel 'Feeding rams') in rams.

If you are concerned that your flock may be suffering from deficiencies, your vet may consider it worth carrying out blood testing.

MINERAL LICKS

If a particular deficiency is diagnosed, then blocks, buckets, or 'licks' can be bought, which contain the required 'top up' nutrients. They are placed in the field for sheep to help themselves and are designed to be weather resistant. High-energy feed blocks can be a convenient way of supplying minerals, especially at times when the available grass is less than ideal. Most sheep find them attractive, but inevitably there will be some that don't take to them. If that is the case, ask your vet about alternative ways of providing supplements.

Some common problems in ewes and lambs

Hypocalcaemia (milk fever; lambing sickness)

- **Cause:** Swift and significant reduction in calcium in pregnant ewes (particularly older ones) – caused by demands of the growing lamb on the mother's body.
- **Symptoms:** Ewes can be nervous, unsteady on their feet, and may be trembling or drooling. They may lie on their bellies with heads down and hind legs stretched out behind. Breathing is likely to be shallow. Ewes become comatose and death can be rapid.
- **Treatment:** Calcium borogluconate solution (e.g. Calciject) given intravenously (i.e. into a vein) will often get ewes back on their feet within minutes, but you have to be sufficiently skilled in order to perform this task. It can be given subcutaneously but may take up to four hours to work.
- **Prevention:** Provide concentrates with good levels of calcium (5–10g per day) alongside forage during pregnancy – at least six to eight weeks prior to lambing.

Hypomagnesaemia (grass staggers)

- **Cause:** Shortage of magnesium. Can affect ewes from lambing time to peak lactation.
- **Symptoms:** Stiffness when walking, facial tremors, frequent urination, rigidity when lying down. Followed by collapse, convulsions and death. Often happens so quickly, sheep may be found dead before action can be taken.
- **Treatment:** Magnesium sulphate and calcium boroglutonate, injected subcutaneously over several sites.
- **Prevention:** If deficiencies are known, treat with magnesium supplement in feed or via bolus. Clover has a high magnesium content compared to grass, so consider introducing more into the sward.

Nutritional muscular dystrophy (white muscle disease)

- **Cause:** Low levels of selenium or vitamin E during pregnancy affect lamb development.
- **Symptoms:** Lambs born dead or weak. Those born live show poor coordination and trembling.
- **Treatment:** Selenium and vitamin E by injection or bolus.
- **Prevention:** Provide ewes with selenium and vitamin E supplements, particularly when pregnant.

Swayback

- **Cause:** Deficiency of copper affects lambs' nervous systems.
- **Symptoms:** Lambs born dead, too weak to stand, or lacking coordination.
- **Treatment:** No suitable treatment; damage is irreversible.
- **Prevention:** Correct deficiencies in ewes before mating by injection or bolus.

FEEDING RAMS

Much is discussed about maintaining the condition of ewes and lambs, but don't forget that your ram needs some special attention, too. NEVER feed rams any concentrates that are specifically designed for ewes, as they tend to have high levels of magnesium and phosphorous, which can contribute to a condition called urinary calculi or urolithiasis (also commonly called 'stones', 'gravel', or 'water belly'). Mineral crystals cause a blockage in the urethra, preventing urination and sometimes leading to rupture of the urethra or the bladder.

If you do need to give a concentrate as a supplement, specialist ram feeds are available, as are general-purpose sheep mixes (e.g. coarse mix), suitable for both ewes and rams.

Poisoning

Sheep aren't too discriminate in what they eat, so beware of giving them access to toxic plants. The main ones to keep them from are yew, bracken and rhododendron. Yew will kill a sheep and there is no cure; bracken will cause a degeneration of the retina, which eventually leads to blindness – and again, there is no cure; rhododendron can cause death if eaten in large quantities but is more likely to cause vomiting.

Copper poisoning

Just as mineral deficiencies can lead to serious health problems, so can consuming too much in some cases. The most common form of mineral overdose involves copper. Chronic copper poisoning can occur in sheep eating large quantities of concentrates – or if other livestock feed (which tend to have higher quantities of copper included) are given. Grazing sheep on pasture previously used by pigs is not recommended because of the high copper content in pig feed.

Copper poisoning damages the intestines, causing abdominal pain, salivation and diarrhoea. In mild cases, sodium sulphate or yellow prussiate of potassium may help, but if the condition is severe, it can be fatal.

Feeding for breeding

As mentioned earlier, both ewes and rams need to be in optimum health in the run-up to the breeding season if they are to perform well. Breeding is discussed in more detail in Chapter 9, but it's worth mentioning at this stage the importance of preparing the future parents for the job ahead.

Looking after the ram

The ram is 'half your flock', so the saying goes. The fact is that if he isn't fit enough to do his job properly, you're going to be disappointed, come lambing time, so pay as much attention to him during the run-up to mating as you do your ewes.

As shown in the table on page 86, you should aim to get rams into condition score 3.5 prior to mating. Good nutrition – along with trace minerals, particularly zinc and selenium – is vital for the production of quality sperm. Sperm takes seven weeks to produce, so make sure that, if he needs feeding up, you start in plenty of time. At least

Richard Stanton Photographic

two months before mating, a ram lacking in condition should be given forage along with a small amount – around 0.2kg a day – of high-protein concentrate to help reach the 3.5 target in time.

Maintaining ram condition throughout the mating period is extremely important, particularly if he is expected to serve large numbers of ewes. Increased periods of exercise and being distracted from normal grazing can take its toll, with a detrimental impact on his performance. When the mating season is over, don't forget to check the ram's condition, as he may have lost weight – in which case he could need additional feeding to prepare him for the winter.

PREGNANCY TOXAEMIA

Pregnancy toxaemia – also known as twin lamb disease – is one of the main problems that can arise as a result of inappropriate feeding during pregnancy. It's most frequently seen in malnourished ewes carrying multiple lambs, but it can also occur in overweight ewes. As the growing lambs make greater demands on the under-nourished ewe's body, her liver breaks down her fat reserves into ketones, which can poison her system. Unfortunately, the signs are often spotted too late for the ewe to be saved. She is likely to avoid the rest of the flock, even at feeding time, and she may also be blind. Grinding of the teeth and frothing of the mouth may be seen, but the classic sign is a smell of acetone or 'pear drops' on the breath. Intravenous glucose injections are the favoured treatment for quick absorption, but there are oral alternatives available.

Care of the ewe

Studies have shown that ewes in good condition at mating time will ovulate better, increasing the likelihood of twins or triplets. On the other hand, poor body condition can delay ovulation and reduce the chance of a successful pregnancy. Recent research suggests that even six months of poor nutrition (e.g. in the case of a very dry summer with poor grass growth) need not have an adverse impact on breeding, as long as additional feed is given in the fortnight before mating. However, farmers have long maintained that upping the feed for ewes from six to eight weeks prior to lambing is the best course of action – with special attention paid to feeding in the final four weeks of pregnancy, when unborn lambs are developing most rapidly. As lambs gain around 70% of their weight during this final month, it's essential to keep the ewe well nourished to maintain condition and avoid problems such as pregnancy toxaemia – also known as 'twin lamb disease' (see panel – left).

The process of increasing feed in order to maximise fertility is known as 'flushing' or 'steaming up'. This can be done either by moving ewes to much better grazing or by supplementary feeding two to three weeks before mating.

Thinner ewes will benefit greatly from flushing, but there is no evidence to suggest there will be any improvement in ewes that are already in good condition. One school of thought is that, if you keep your ewes in good condition at other times of the year, they shouldn't need flushing. If their condition is below what's required, start by supplementing forage with ewe nuts at the rate of 250g a day, six weeks before lambing, increasing gradually week by week, up to 1kg a day. However, keep checking condition to avoid overfeeding. Be particularly careful that you don't overfeed young ewes, shearlings, or ewe lambs lambing for the first time, in order to avoid oversized lambs and difficult births.

Remember, at all times, that you are aiming for fit rather than fat – in the case of both ewes and rams.

BREEDING

Think before breeding

Have a good think before you commit yourself to becoming a sheep breeder.

- Do you really want all the hassle and the hard work?
- Do you have a suitable building where your ewes can lamb? If not, are you ready for outdoor lambing and everything that comes with it – including round-the-clock stock checks in all weathers?
- Could you cope with the more gory side of the birthing process?
- Are you ready to deal with sickly lambs, deformities, birthing complications and deaths?
- Do you have a market for any lambs produced?
- Do you have sufficient grazing to support not only your existing sheep but their offspring too?

If you can answer 'yes' to all these questions, then read on. Just don't leap too early into a situation for which you may not be prepared. Take time to get to know your sheep and their needs first.

Pre-mating checks

As with all livestock, only the best animals should be used as breeding stock, so make sure you have carried out all the basic checks (condition scoring and, of course, the '3 Ts'!) mentioned earlier to confirm that your sheep are up to standard and don't have any genetic defects which could be passed on. Don't make the mistake of breeding from anything that doesn't make the grade: breed from the best and get rid of the rest!

Overall health, fitness, and mobility assessments should also be carried out at this time, as should any relevant vaccinations (see Chapter 6).

Selecting a ram

If you only have a few ewes, it won't be practical to keep your own ram, so you will probably end up borrowing one from a neighbouring breeder. Generally, a ratio of one ram to every 30–50 ewes is used in lowland flocks, while on hill farms it can be one for 80–100. If you decide to buy a ram of your own, there are lots of factors to bear in mind about selection. Look back at Chapters 4 and 8 for a reminder of what to aim for as far as the overall structure and physical features are concerned, as you need to make sure that any ram you use is sufficiently suited for the task ahead.

WHO'S THE DADDY?

Just because a ram looks like a ram, don't automatically assume he will be able to sire lambs. Buying an unproven ram (one that has not yet produced offspring) can end in disappointment. Sometimes a young ram lacks experience and technique; he may have been mismatched in size to the ewes; or he may simply be infertile or not willing to work. Research work carried out in the United States by Dr Charles E. Roselli of the Oregon Health and Science University has revealed that up to 10% of rams are homosexual and will not mate with ewes – although they may mount other males.

In large commercial operations, a breeding soundness examination – a physical assessment and a semen test – may be carried out on new rams.

PEDIGREE BREEDING – IS IT WORTH IT?

Now there's a tricky question! If all you want to do is to produce strong, healthy stock that will provide you and your customers with excellent meat, you may think there is little point in going to the expense of buying pedigree stock. So why bother buying a sheep that comes with a piece of paper detailing its family tree but costs several times more than one that looks identical but has no documented ancestry?

There are lots of reasons, really. On the face of it, the world of pedigree sheep can be a lucrative one, and you only have to look at the staggering prices raised at auction to be tempted. The world's most expensive sheep to date – Deveronvale Perfection, a Texel ram lamb, – was sold in Lanark, Scotland, in 2009 for £231,000. Previously, the highest price paid for a sheep was £205,000, for a Merino ram sold in 1989 in Australia, although more recently another British-bred Texel ram lamb made a respectable £152,000 in 2014. There's no doubt that the new owners of these sheep will see their investment multiplied several times over by hiring them out to serve other flocks or selling offspring and semen worldwide.

Of course, these sorts of prices – the kind that could buy you a comfortable family home in many places – are at the extreme end of the market. Producing top-quality commercial rams and ewes to be used for artificial insemination and fertilised embryos is a highly specialised market, and not one into which newcomers can step overnight.

Having said that, it is possible to get a foothold by buying good foundation stock, and having the ability to sell registered stock potentially increases the number of outlets you have for your lambs. Rare breeds – as well as some fashionable or more 'novelty' breeds (pictured here) – can be a niche market, depending on where you live, what people are prepared to pay and, ultimately, how good your sheep are. Quality and correctness are crucial. It may take many years to build up the knowledge and expertise required to turn out good examples of the breed – and even longer to build a good reputation.

As mentioned earlier, each breed has its own 'standard' to ensure the quality of animals being registered in the official flock book. With some breeds, sheep can only be registered (some societies say 'approved') and admitted to the flock book if they have been inspected by a society official. Inspections are often carried out at regional venues on specified dates, while some societies will visit individual farms to examine stock. All breeds have different selection criteria, so make sure you understand what is acceptable and what is not before you make a purchase. Visit agricultural shows and pedigree sheep sales and talk to the exhibitors about what makes a good example of the breed. Most are keen to get newcomers involved and will be happy to give advice.

In addition to the statutory tagging obligations, identification for pedigree breeding may also involve additional tagging, tattooing or notching. Check with the relevant breed society for regulations, as there will be variations.

Supporting rare breeds

Conservation is one reason why many people decide to keep native rare breeds. The Rare Breeds Survival Trust publishes an annual 'Watchlist' of the most endangered breeds in the UK. They are split into categories according to their numbers, namely Critical, Endangered, Vulnerable, At Risk and Minority – with those in the 'Critical' category giving most cause for concern.

The lists are compiled using information about the total number of registered breeding females. Other factors are also taken into consideration, including geographical concentration – for instance, where a large percentage of the breed are found in a relatively small area. If you turn to the Guide to sheep breeds (page 152) the letter 'R' appears next to the name of any breed on the Watchlist at the time of writing (2015).

Emma Collison

Above and below: Fashionable breeds like the Valais Blacknose (top) and the diminutive Babydoll (bottom) can be money-spinners.

When choosing a ram you should be even more cautious and critical than when choosing ewes, and you should also be prepared to pay a fair sum for a good one. As the ram's genetics will have a major impact on so many offspring, you need to consider not only the obvious things like appearance and bodily attributes but also the genetic improvements he might be able to make. For instance, a ram which was born as one of twins or triplets, or which came from a particularly prolific (i.e. productive) ewe, has the potential to pass on this breeding strength to his offspring.

In the case of pedigree flocks, the choice of breeding stock is even more difficult, as both parents must meet the breed standard – the 'checklist' of essential or desirable features specified by the breed society.

If you are aiming to produce lambs to sell at market rather than solely for your own consumption, you may want to choose a well-grown and well-muscled ram to produce lambs with good, meaty carcasses. However, be aware that some large-framed rams, when mated with ewes of smaller breeds, may sire large lambs, which are difficult to deliver, so take advice on what might be the most suitable cross.

Choosing your ewes

There are a few different choices to be made when it comes to buying your female breeding stock. You can start with ewe lambs or yearlings/shearlings, which have no experience of lambing, or you can buy older ewes, which have already lambed and know the ropes; you can buy 'empty' ewes and aim to get them in lamb yourself, or buy some which are already pregnant.

FERTILITY TESTING

A fine pair of testicles and good overall body shape doesn't automatically mean that a ram will be a good breeder. Sometimes even the best-looking rams turn out to be infertile – 'firing blanks', as they say. Your vet can carry out a full examination – a physical assessment of the penis, prepuce, scrotum and epididymis, followed by a semen test – for a relatively small cost.

Occasionally there is no physical sign as to why a ram should not want to mate; some just won't do the job, or choose to serve infrequently – often causing an extended lambing season. Over-fat or undernourished rams may show poor sex drive, and lack of libido can also be an inherited trait. Keenness to mate can also be diminished by age and health problems.

Ewe lambs: what age to breed?

Shepherds are divided on this one: there are some who will breed from a ewe lamb as soon as she is sufficiently well-grown; others will wait until she has been through her first year and has been shorn (known as a 'yearling' or 'shearling') before putting her to the ram.

The economics of sheep keeping would seem to suggest that producing new lambs as early as possible is a good idea. However, early spring ewe lambs will still be growing by mating time in the autumn and, as their bodies are still developing, they may have problems delivering and need assistance. Multiple births are not unknown in ewe lambs and, occasionally, they may not produce enough milk to feed their offspring. For these reasons, it's probably best to go with yearlings or older

Young stock v older stock

	Pros	Cons
Ewe lambs/ yearlings	Younger stock grow up with you and so are easier to tame, making for easier handling – particularly at lambing time.	Buying 'unproven' stock means there is no guarantee of fertility; and, if they do get pregnant, there is no way of knowing what kind of deliveries they might have, nor how they will perform as mothers.
	You are less likely to be buying in problems, e.g. damaged udders, recurring prolapse, poor teeth.	As they have more productive years ahead of them, prices will be higher than for older ewes.
Older ewes	Some of the worry is removed, as they have been through the lambing process before – even if you haven't. Dealing with a reputable breeder, you should be able to buy older ewes with confidence.	If you buy at auction and don't know the vendor, you could be taking on troublesome beasts that he or she is keen to offload because of previous problems.
	They will be cheaper to buy, and you should get a few crops of lambs from them before they need culling or require extra attention. Experienced farmers often buy older ewes cheaply at market with the knowledge that they will be able to produce some good replacement stock.	Won't necessarily last as long as younger ewes, plus fertility and ease of lambing can decline with age. Teeth may be worn or missing, so maintaining condition could be a problem.

ewes first time round. Aim to make your first lambing season as stress-free as possible!

There's a lot to be said for choosing a ewe that has been through one or two successful lambings. As well as knowing what they're supposed to do, older ewes are more likely to have multiple lambs, they tend to be better mothers, and normally produce plenty of milk.

If you decide to put ewe lambs to the ram, there are a few considerations to bear in mind. It's been shown that well-nourished, well-grown lambs can be fit enough to breed in their first year of life and give birth by their first birthday – as long as their weight at mating is at least 60% of the mature body weight of the breed. They should be kept separately from more mature ewes at mating time and afterwards, in order to keep a close eye on feeding and maintaining condition. Careful feeding is essential: ewe lambs have been found to require 20% more feed than mature ewes in order to sustain their own continuing body growth as well as supporting unborn lambs. Their offspring should be weaned earlier than other lambs; they should be creep-fed prior to weaning and separated as early as eight weeks old, to avoid placing too much demand on the body of the young mother.

Planning your lambing season

The majority of British breeds are seasonal or 'short day' breeders, which means they are programmed to be receptive to mating during the late autumn and winter months, when the days shorten. There are a few exceptions to the rule, including Dorset (both polled and horned), Portland and Merino, which can breed, pretty much, at any time of the

year. In seasonal breeders, preparation for mating is kick-started by the increase in the hours of darkness. Reduced daylight stimulates the release of the hormone melatonin. This triggers a series of other biological reactions, which lead to the onset of oestrus – the period when ewes are sexually receptive and fertile. During this time, the ewe is often described as being 'in season' or 'on heat'.

Opting for a non-seasonal breed has its advantages. Ewes can be mated in batches to lamb at different times of the year, creating a constant supply of lambs and easing the shepherd's workload. There is also the opportunity to sell new-season lamb while other farmers are still waiting for their later-born lambs to grow. There's no doubt that supervising an outdoor-lambing flock during late spring and early summer is a much more enjoyable activity for the breeder!

SPRING LAMBS

The major benefit of having sheep coming into season towards the tail end of the year is survivability of lambs. The gestation period for sheep lasts between 144 and 152 days; therefore ewes that are mated in October and November, for instance, will be born in March and April – a time when the weather tends to be kinder and grass growth is improving. Generations of farmers have traditionally aimed for the lambing season to start in the spring but, following a series of wetter winters, an increasing number of flocks – particularly those lambing outdoors – are being held back to lamb in May, to give lambs a better start in life and also to minimise feed costs because grazing is more plentiful.

SHEEP GESTATION GUIDE

Date of Service	Watch for Lambs	Date of Service	Watch for Lambs	Date of Service	Watch for Lambs	Date of Service	Watch for Lambs	Date of Service	Watch for Lambs	Date of Service	Watch for Lambs	Date of Service	Watch for Lambs
Jan 1	May 26	Feb 24	Jul 19	Apr 19	Sep 11	Jun 12	Nov 4	Aug 5	Dec 28	Dec 28	Sep 28	Nov 21	Apr 15
Jan 2	May 27	Feb 25	Jul 20	Apr 20	Sep 12	Jun 13	Nov 5	Aug 6	Dec 29	Dec 29	Sep 29	Nov 22	Apr 16
Jan 3	May 28	Feb 26	Jul 21	Apr 21	Sep 13	Jun 14	Nov 6	Aug 7	Dec 30	Dec 30	Sep 30	Nov 23	Apr 17
Jan 4	May 29	Feb 27	Jul 22	Apr 22	Sep 14	Jun 15	Nov 7	Aug 8	Dec 31	Dec 31	Oct 01	Nov 24	Apr 18
Jan 5	May 30	Feb 28	Jul 23	Apr 23	Sep 15	Jun 16	Nov 8	Aug 9	Jan 1	Jan 1	Oct 2	Nov 25	Apr 19
Jan 6	May 31	Mar 1	Jul 24	Apr 24	Sep 16	Jun 17	Nov 9	Aug 10	Jan 2	Jan 2	Oct 3	Nov 26	Apr 20
Jan 7	Jun 1	Mar 2	Jul 25	Apr 25	Sep 17	Jun 18	Nov 10	Aug 11	Jan 3	Jan 3	Oct 4	Nov 27	Apr 21
Jan 8	Jun 2	Mar 3	Jul 26	Apr 26	Sep 18	Jun 19	Nov 11	Aug 12	Jan 4	Jan 4	Oct 5	Nov 28	Apr 22
Jan 9	Jun 3	Mar 4	Jul 27	Apr 27	Sep 19	Jun 20	Nov 12	Aug 13	Jan 5	Jan 5	Oct 6	Nov 29	Apr 23
Jan 10	Jun 4	Mar 5	Jul 28	Apr 28	Sep 20	Jun 21	Nov 13	Aug 14	Jan 6	Jan 6	Oct 7	Nov 30	Apr 24
Jan 11	Jun 5	Mar 6	Jul 29	Apr 29	Sep 21	Jun 22	Nov 14	Aug 15	Jan 7	Jan 7	Oct 8	Dec 1	Apr 25
Jan 12	Jun 6	Mar 7	Jul 30	Apr 30	Sep 22	Jun 23	Nov 15	Aug 16	Jan 8	Jan 8	Oct 9	Dec 2	Apr 26
Jan 13	Jun 7	Mar 8	Jul 31	May 1	Sep 23	Jun 24	Nov 16	Aug 17	Jan 9	Jan 9	Oct 10	Dec 3	Apr 27
Jan 14	Jun 8	Mar 9	Aug 1	May 2	Sep 24	Jun 25	Nov 17	Aug 18	Jan 10	Jan 10	Oct 11	Dec 4	Apr 28
Jan 15	Jun 9	Mar 10	Aug 2	May 3	Sep 25	Jun 26	Nov 18	Aug 19	Jan 11	Jan 11	Oct 12	Dec 5	Apr 29
Jan 16	Jun 10	Mar 11	Aug 3	May 4	Sep 26	Jun 27	Nov 19	Aug 20	Jan 12	Jan 12	Oct 13	Dec 6	Apr 30
Jan 17	Jun 11	Mar 12	Aug 4	May 5	Sep 27	Jun 28	Nov 20	Aug 21	Jan 13	Jan 13	Oct 14	Dec 7	May 1
Jan 18	Jun 12	Mar 13	Aug 5	May 6	Sep 28	Jun 29	Nov 21	Aug 22	Jan 14	Jan 14	Oct 15	Dec 8	May 2
Jan 19	Jun 13	Mar 14	Aug 6	May 7	Sep 29	Jun 30	Nov 22	Aug 23	Jan 15	Jan 15	Oct 16	Dec 9	May 3
Jan 20	Jun 14	Mar 15	Aug 7	May 8	Sep 30	Jul 1	Nov 23	Aug 24	Jan 16	Jan 16	Oct 17	Dec 10	May 4
Jan 21	Jun 15	Mar 16	Aug 8	May 9	Oct 1	Jul 2	Nov 24	Aug 25	Jan 17	Jan 17	Oct 18	Dec 11	May 5
Jan 22	Jun 16	Mar 17	Aug 9	May 10	Oct 2	Jul 3	Nov 25	Aug 26	Jan 18	Jan 18	Oct 19	Dec 12	May 6
Jan 23	Jun 17	Mar 18	Aug 10	May 11	Oct 3	Jul 4	Nov 26	Aug 27	Jan 19	Jan 19	Oct 20	Dec 13	May 7
Jan 24	Jun 18	Mar 19	Aug 11	May 12	Oct 4	Jul 5	Nov 27	Aug 28	Jan 20	Jan 20	Oct 21	Dec 14	May 8
Jan 25	Jun 19	Mar 20	Aug 12	May 13	Oct 5	Jul 6	Nov 28	Aug 29	Jan 21	Jan 21	Oct 22	Dec 15	May 9
Jan 26	Jun 20	Mar 21	Aug 13	May 14	Oct 6	Jul 7	Nov 29	Aug 30	Jan 22	Jan 22	Oct 23	Dec 16	May 10
Jan 27	Jun 21	Mar 22	Aug 14	May 15	Oct 7	Jul 8	Nov 30	Aug 31	Jan 23	Jan 23	Oct 24	Dec 17	May 11
Jan 28	Jun 22	Mar 23	Aug 15	May 16	Oct 8	Jul 9	Dec 1	Sept 1	Jan 24	Jan 24	Oct 25	Dec 18	May 12
Jan 29	Jun 23	Mar 24	Aug 16	May 17	Oct 9	Jul 10	Dec 2	Sept 2	Jan 25	Jan 25	Oct 26	Dec 19	May 13
Jan 30	Jun 24	Mar 25	Aug 17	May 18	Oct 10	Jul 11	Dec 3	Sept 3	Jan 26	Jan 26	Oct 27	Dec 20	May 14
Jan 31	Jun 25	Mar 26	Aug 18	May 19	Oct 11	Jul 12	Dec 4	Sept 4	Jan 27	Jan 27	Oct 28	Dec 21	May 15
Feb 1	Jun 26	Mar 27	Aug 19	May 20	Oct 12	Jul 13	Dec 5	Sept 5	Jan 28	Jan 28	Oct 29	Dec 22	May 16
Feb 2	Jun 27	Mar 28	Aug 20	May 21	Oct 13	Jul 14	Dec 6	Sept 6	Jan 29	Jan 29	Oct 30	Dec 23	May 17
Feb 3	Jun 28	Mar 29	Aug 21	May 22	Oct 14	Jul 15	Dec 7	Sept 7	Jan 30	Jan 30	Oct 31	Dec 24	May 18
Feb 4	Jun 29	Mar 30	Aug 22	May 23	Oct 15	Jul 16	Dec 8	Sept 8	Jan 31	Jan 31	Nov 1	Dec 25	May 19
Feb 5	Jun 30	Mar 31	Aug 23	May 24	Oct 16	Jul 17	Dec 9	Sept 9	Feb 1	Feb 1	Nov 2	Dec 26	May 20
Feb 6	Jul 1	Apr 1	Aug 24	May 25	Oct 17	Jul 18	Dec 10	Sept 10	Feb 2	Feb 2	Nov 3	Dec 27	May 21
Feb 7	Jul 2	Apr 2	Aug 25	May 26	Oct 18	Jul 19	Dec 11	Sept 11	Feb 3	Feb 3	Nov 4	Dec 28	May 22
Feb 8	Jul 3	Apr 3	Aug 26	May 27	Oct 19	Jul 20	Dec 12	Sept 12	Feb 4	Feb 4	Nov 5	Dec 29	May 23
Feb 9	Jul 4	Apr 4	Aug 27	May 28	Oct 20	Jul 21	Dec 13	Sept 13	Feb 5	Feb 5	Nov 6	Dec 30	May 24
Feb 10	Jul 5	Apr 5	Aug 28	May 29	Oct 21	Jul 22	Dec 14	Sept 14	Feb 6	Feb 6	Nov 7	Dec 31	May 25
Feb 11	Jul 6	Apr 6	Aug 29	May 30	Oct 22	Jul 23	Dec 15	Sept 15	Feb 7	Feb 7	Nov 8		
Feb 12	Jul 7	Apr 7	Aug 30	May 31	Oct 23	Jul 24	Dec 16	Sept 16	Feb 8	Feb 8	Nov 9		
Feb 13	Jul 8	Apr 8	Aug 31	Jun 1	Oct 24	Jul 25	Dec 17	Sept 17	Feb 9	Feb 9	Nov 10		
Feb 14	Jul 9	Apr 9	Sep 1	Jun 2	Oct 25	Jul 26	Dec 18	Sept 18	Feb 10	Feb 10	Nov 11		
Feb 15	Jul 10	Apr 10	Sep 2	Jun 3	Oct 26	Jul 27	Dec 19	Sept 19	Feb 11	Feb 11	Nov 12		
Feb 16	Jul 11	Apr 11	Sep 3	Jun 4	Oct 27	Jul 28	Dec 20	Sept 20	Feb 12	Feb 12	Nov 13		
Feb 17	Jul 12	Apr 12	Sep 4	Jun 5	Oct 28	Jul 29	Dec 22	Sept 21	Feb 13	Feb 13	Nov 14		
Feb 18	Jul 13	Apr 13	Sep 5	Jun 6	Oct 29	Jul 30	Dec 23	Sept 22	Feb 14	Feb 14	Nov 15		
Feb 19	Jul 14	Apr 14	Sep 6	Jun 7	Oct 30	Jul 31	Dec 24	Sept 23	Feb 15	Feb 15	Nov 16		
Feb 20	Jul 15	Apr 15	Sep 7	Jun 8	Oct 31	Aug 1	Dec 25	Sept 24	Feb 16	Feb 16	Nov 17		
Feb 21	Jul 16	Apr 16	Sep 8	Jun 9	Nov 1	Aug 2	Dec 26	Sept 25	Feb 17	Feb 17	Nov 18		
Feb 22	Jul 17	Apr 17	Sep 9	Jun 10	Nov 2	Aug 3	Dec 27	Sept 26	Feb 18	Feb 18	Nov 19		
Feb 23	Jul 18	Apr 18	Sep 10	Jun 11	Nov 3	Aug 4	Dec 28	Sept 27	Feb 19	Feb 19	Nov 20		

This gestation chart will help you work out when your lambs are due. Make a note of the date you introduce the ram to your ewes – or, even better, the date you see them being served and look for it on the chart. The approximate due date will be immediately below. This chart is based on a 145-day gestation. However, there will be differences between breeds, and factors like stress and environmental conditions can also have an impact on due dates.

Above: A ram showing the 'flehmen response'.

jane Bissett

Understanding how breeding works

The ewe's fertility and willingness to mate is governed by the oestrus cycle, which can be between 13 and 19 days, depending on breed, with the average being 17. The time she will stand to be mated lasts just 24 to 36 hours, so there's just a small 'window of opportunity' for the ram to get the job done.

Unlike some other mammals, ewes don't often show obvious signs of being in season. Although older ewes will sometimes actively go looking for the ram and stand to be mounted – or even nudge him, demanding sex – the real give-away is the behaviour of the ram. Sniffing of the ewes' genitals and curling back the top lip (known as 'the flehmen response') are classic signs of being interested. When in the presence of ewes, he will give off pheromones, which prompt sexual receptiveness in the flock. If the ewes have been kept from seeing and smelling the ram for three weeks, they should ovulate within two to four days of coming into contact with him.

Many sheep farmers prefer not to have a long, drawn-out lambing season, and there are various ways of encouraging ewes to lamb closer together and/or earlier than nature would normally allow:

Below: Ram sniffing a ewe in season.

Using a 'teaser'

A teaser is a vasectomised ram, which is used to get the ewes 'in the mood' before the fertile ram is introduced and to minimise the time between each ewe giving birth. Vasectomy should only be carried out by a veterinary surgeon, and it can be done on site, under local anaesthetic and with sedation. Even though he is unable to fertilise eggs, the teaser will still stimulate oestrus in the ewes. This is known as the 'ram effect'.

Ewes should not have contact with any rams – vasectomised or entire – for at least three weeks before the teaser is introduced. He should stay with the flock for up to 14 days and then be taken out and swapped with the intact ram.

'Sponging'

A sponge containing a synthetic, progesterone-like hormone is inserted into the vagina for 12 to 14 days. The sponge – which is similar to a woman's sanitary tampon, with a cylindrical shape and a string for easy removal (see photos below) convinces the body it is pregnant, preventing ovulation. When it is removed, the ewe will normally ovulate within 36 to 72 hours. It's recommended that rams are introduced between 36 and 48 hours after the sponges have been removed, and that they should be kept there for at least 48 hours. This system works fairly well during the

1 A sponge is inserted using an applicator
2 The applicator is withdrawn
3 The sponge is in position – with strings visible
4 When the correct number of days has passed, the sponge is withdrawn

MSD Animal Heath

natural breeding season, but if an early lambing season is required – such as December or January – an injection of PMSG (pregnant mare's serum gonadotropin) should be given at the time of sponge removal.

As well as enabling the shepherd to choose when lambing starts, sponging also means that ewes can be synchronised to give birth in one group or in batches to suit individual requirements.

SPONGING TIMETABLE

- Decide on the start of your lambing season when you would like to start lambing.
- Subtract 147 days to give you the date the ram is introduced.
- Subtract another two days (to account for the 48-hour waiting period). This gives you the date the sponges are taken out.
- Subtract another 14 days. This gives you the date the sponges are inserted.

A few things need to be borne in mind when using the sponging technique, as this system puts more pressure on the ram, which will be expected to serve far more ewes in a shorter space of time than normal. For this reason, you should always use a proven ram rather than an untested one or a ram lamb. He should always be in top condition, and the breeding ratio should be one ram to no more than 10 ewes. If you are aiming for an earlier-than-normal lambing, you should be aware that the quality of the ram's semen is not at its best outside the normal breeding season, so you may need to use one ram to five ewes.

Melatonin implants

As touched on earlier, melatonin is the hormone released from the pineal gland in the brain as hours of daylight shorten and is the natural starting point in the breeding process. Implants (e.g. Regulin) are available, which artificially increase the levels in the body and can bring the breeding season forward by as much as eight weeks. The small, biodegradable implant is injected under the skin at the base of the ear and induces oestrus 50 to 70 days later. Implants can also be used in rams to increase libido and improve the quantity and quality of sperm.

Using a raddle

Rather than watch their flocks like a hawk for signs of mating activity, many breeders prefer to let the rams do the work for them. The easiest way of spotting if a ram has mounted a ewe is to put a coloured raddle marker on his chest, which is transferred to the ewe on contact. This procedure is known as 'raddling'. Raddles can also be used when different rams let loose in the same flock, as a way to identify which ones are working, and on which ewes.

The ram is fitted with a harness holding a wax crayon to colour the ewe's rump – or, a more low-tech approach is simply to smear raddle paste on to his brisket. Raddle powder can be bought from agricultural suppliers and mixed with vegetable oil to make a thick paste. The one downside is that the ram will have to be caught to have the paste reapplied; crayons last longer. If a harness is used, it should be fitted two days before the ram is due to start work, to allow him to get used to it. It should also be checked each day to make sure it is not rubbing the ram in an

ARTIFICIAL INSEMINATION

Artificially inseminating ewes is yet another way of organising your breeding season, as well as giving an opportunity to introduce some really good genetics into your flock at a fraction of the cost of investing in a top-class ram. There are two methods: cervical and laparoscopic. The first involves introducing semen directly into the uterine horn. Each procedure can only be carried out if ewes are in season, therefore sponging needs to be carried out beforehand and semen ordered accordingly.

Cervical AI is a reasonably cheap procedure – costing just a few pounds per ewe – but it does require at least three people, as the ewe needs to be well restrained by two pairs of hands while she is inseminated.

Laparoscopic AI is a much more complex – and costly – surgical technique, which can only be carried out by a veterinarian or a suitably skilled technician. It involves sedating the ewes, holding them upside down in a crate, and making an incision into the abdomen wall.

Embryo transfer

Fertilised embryos can be taken from a donor ewe and transplanted into 'surrogate' ewes. Hormone treatment is given to the donor ewe in order to produce more eggs than normal and then AI is carried out. Once fertilised, the eggs can either be transferred straight into another ewe or frozen for future use.

Sourcing semen

A quick Internet search will reveal there are what appear to be 'dating agencies' for sheep! Websites like www.rams4ewe.co.uk provide opportunities to view photographs of rams of numerous breeds, look up their pedigree names, and do your research. Some breed societies also offer pedigree semen for sale. Fresh semen produces better results, but frozen semen can be stored for many years, enabling it to be used long after the donor ram has died – a factor particularly important in respect of rare breeds and scarce bloodlines.

Smearing paint on to the brisket.

Jan Walton

A Dorset ram wearing a raddle.

David and Ruth Wilkins, Rampisham Farm

uncomfortable way, which may put him off his job. Another thing to check is that it is not too big and not too small, as rams may lose weight through exercise or fatten up when put on good pasture.

Start with a light-coloured crayon or paste so that you can move on to a darker one and still see the difference. Shepherds have differing opinions on the frequency raddle colours should be changed. Some prefer to do it roughly halfway through the 17-day oestrus cycle (8 to 10 days), and again at day 17; others stick to every 17 days. If a ewe is marked first time round, it can be assumed that she has been inseminated, and the approximate mating date noted – but, of course, there is no guarantee that a pregnancy will hold. If she is marked twice, make a note of the second approximate mating date and work out the likely lambing dates using the gestation chart on page 98.

Post-mating management

The early weeks of pregnancy are crucial for the successful implantation and retention of fertilised eggs. Feeding should be maintained at the same level as prior to mating and every effort should be made to avoid stressful situations around mating time and afterwards. If ewes are anxious, pregnancy is less likely to occur. Minimum intervention and careful handling is essential. It takes around three days for a fertilised egg (now called an embryo) to make its way through the oviduct and enter the uterus. Day 15 to day 30 of pregnancy, when the embryo attaches to the lining of the uterus, is a sensitive period, during which many pregnancies (as many as 25%) fail. The ewe will normally then come back into season – often described as 'returning to service'.

Scanning ewes
Getting an ultrasound scan of your ewes will give you peace of mind for several reasons. Firstly, you will know whether they are pregnant or not! If any are not, you may still have time to put them back to the ram, and you may also want to get blood tests carried out to rule out any medical cause of infertility, such as disease. Scanning tells you how many lambs each ewe is carrying, so you can work out how much they need to eat to

support them. As mentioned in Chapter 8, ensuring that your ewes are getting the nutrition they need through pregnancy is essential. Scanning also gives you a clearer picture of what kind of lambing to expect, so you can plan ahead. If you know that a ewe is having twins and you have to assist her during lambing, you will know how many sets of legs you need to be feeling for. Another advantage of scanning is that you can identify barren ewes and sell them on as culls straight away, rather than assuming they are pregnant and wasting feed on them.

Personal preferences regarding when to scan vary a great deal – some like to do it soon after the first month of a suspected pregnancy has passed, while others prefer to wait until around three months in. A skilled scanning technician is often able to identify foetuses as early as 30 days but, as there is a risk of losses in the early weeks, it's more common to wait a while longer. Foetuses should be easily visible and possible to count by 50 to 55 days, but it's not unusual for some farmers to wait until 100 days or more.

Depending on where you live, persuading a scanner to come to you may prove difficult if you only have a few sheep. Checking with neighbouring farmers to see when their sheep are being scanned is one way around the problem, or you may be able to get together with some other small-scale sheep keepers in your area to make the scanner's visit more viable. A good scanner will be able to process more than 200 ewes in an hour. Most scanners prefer to work with the ewe standing in a narrow space, so a clearer picture of the uterus can be obtained.

Preparing for the scanner's arrival
■ Get the sheep penned up beforehand, so you don't waste the scanner's time while you round them up.
■ Keep the ewes' stress levels to a minimum and have helpers available to open and close gates and hurdles.
■ The scanner may bring a purpose-built trailer, on to which each ewe will be loaded, but find out in advance what equipment he or she uses, so you can set up a makeshift handling system if you don't already have a permanent set-up.
■ Don't feed ewes too much before scanning, as a full rumen can make diagnosis more difficult.
■ Have marker sprays to hand in order to identify ewes with singles, twins, triplets – maybe even more! Don't forget a notebook to keep a tally of how many lambs are scanned for each ewe, and to record anything unusual.

LAMBING

NOTE: This part of the book is not a stand-alone section, which tells you everything you need to know about lambing! It *must* be read in conjunction with other chapters – particularly A healthy flock, Feeding and nutrition, and Breeding – in order to make sure you are fully prepared.

Lambing can be the most magical time of the year – but also the most nerve-racking and distressing if you aren't completely prepared. Sheep have been giving birth all on their own for thousands of years, mostly without any problems and without human assistance, but you do need to be confident that if your ewes get into difficulties, you will be able to help, or at least recognise a problem and summon veterinary assistance.

If you have followed the advice elsewhere in this book regarding the care and welfare of your flock, there shouldn't be any nasty surprises ahead. Prevention, as the saying goes, is better than cure, so there are some key things to bear in mind when caring for pregnant ewes. Refer back to Chapter 8 to refresh your memory about potential difficulties that can occur as a result of nutritional deficiencies and imbalances – many preventable with proper care. Among them are pregnancy toxaemia, hypocalcaemia and hypomagnesaemia (see pages 89 and 91).

COUNTDOWN TO LAMBING

The final six- to eight-week period before lambing is one of the most crucial times in the shepherd's year. Careful management and forward planning will help minimise the likelihood of problems with ewes, lamb deaths and low birth weights.

Number of weeks before lambing	Tasks
8 weeks	■ If you can, book yourself on a good lambing course or offer to help a neighbouring farmer before you have to deal with your own lambing. Even if you've done it before, a refresher is always worthwhile. ■ Check body condition (see Chapter 8) and feed accordingly, as lambs begin growing rapidly from now on, and udder development will be starting, too. Don't be tempted to feed too much; otherwise you risk large lambs, which can be difficult to deliver. ■ Consider scanning ewes if not already done.
6 weeks	■ Discuss your lambing plans with your vet and take advice on what should be in your lambing kit. A routine visit to check your ewes are in good health will give you peace of mind if this is your first lambing season. ■ If you vaccinate against clostridial diseases and pasturella (see Chapter 6), give the annual booster to protect both ewes and their lambs. ■ If ewes have been scanned, divide them into groups according to the number of lambs expected. If any are underweight, despite being fed the correct rations, consider blood testing for metabolic diseases, nutritional imbalances, or internal parasites.
4 weeks	■ Prepare the lambing shed or outdoor paddocks (see page 106). ■ Assemble your lambing kit (see page 106).
2 weeks	■ Gather any ewes that are to lamb indoors and house them according to the number of lambs expected. If possible, keep younger ewes separate to older ones, to avoid bullying. ■ Check for any lameness before housing and treat any cases of foot rot, which can spread rapidly once indoors. ■ Make sure there is sufficient space for each ewe to feed (45cm per ewe is recommended). Keep water clean and fresh.
1 week	■ Keep a close check on ewes in case of early lambers/miscalculated due dates. ■ If you have a large number to lamb, or if you are away from home a lot, organise a rota of helpers to help with monitoring.

Indoor or outdoor lambing?

This is often a matter of personal preference, but if you have no suitable outbuildings, you may have no choice. There are pros and cons with both systems. Lambing indoors gives you more opportunity to supervise your flock and intervene in the case of problems. However, health problems can spread quickly indoors and ewes that are not used to being handled or confined may be more stressed. Outdoor lambing is obviously more natural and is generally considered healthier – though in extreme cold and wet weather, lambs can die from hypothermia or because of unseen birthing difficulties. There is also the risk of mis-mothering (see page 118) and abandonment, too, both of which can lead to starvation.

Other threats to newborns can include foxes, roaming dogs, crows, magpies, birds of prey – and two-legged predators.

One of the key ways to ensure safer lambing is having control over your ewes, which can be achieved by containing them in a much smaller, more manageable area. Sheep hurdles, fence panels, or even wooden pallets can be used effectively to build a moveable mini-paddock in which to house your ewes prior to lambing, and to create individual pens in which each mother and her offspring can bond without interference from others in the flock. Plastic calf hutches (below) are both practical and easily moved to fresh ground, but will need staking down in high winds. A lambing paddock constructed around an open-fronted barn or field shelter, or against a high wall or hedge will give extra protection from the elements. Polytunnels are also a popular choice for lambing and can prove an economical way of erecting a very versatile, multi-purpose structure.

Below: Indoor lambing pen.

Below: Lambing outdoors using temporary shelters.

LAMBING KIT CHECKLIST

Speak to your vet well in advance for recommendations of drugs and other items to include in your lambing kit. Here are some basics with which to start:

- Disposable latex gloves – short and arm-length
- Lubricating gel
- Navel disinfectant/tincture of iodine
- Navel clamps
- Lambing rope or snare
- Syringes and needles
- Stomach feeding tube
- Colostrum and milk replacer – plus feeding bottles, teats, mixing jug, kettle, fork/whisk
- Infrared heat lamp and a box big enough to hold a lamb
- Prolapse 'spoon'/retainer
- Antibiotics and anti-inflammatories
- Calcium borogluconate (e.g. Calciject)
- Glucose solution
- Thermometer
- Rubber rings and elastrator pliers
- Foot clippers, antibacterial foot spray, and dagging shears
- Head-torch

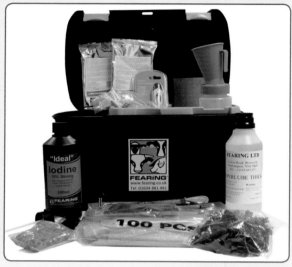

Fearing International www.fearing.co.uk

- Spray-marker
- Notebook
- Vet's contact details
- Mobile phone (charged!)
- Waterproof lamb coats – if lambing early outdoors

The lambing shed

A lambing shed should be well ventilated but free of draughts, and the building should be cleaned and disinfected before use, with bedding replaced regularly. It needs to be large enough to comfortably accommodate the ewes, so make sure they have sufficient space to lie down – at least 1.3m x 1.3m (roughly 4ft x 4ft) and that they can easily get to the feed and water troughs.

Ewes can be kept in small groups prior to lambing, but it's best to move them to their own quarters afterwards, to allow lambs to suckle successfully without interruption and to bond with their mothers. A pen at least 2m x 1m is sufficient as a short-term 'nursery'. Each pen should have a hay rack, along with feed and water containers. Make sure the hay rack is safe and secure, to avoid the ewe dislodging it during feeding, and use shallow water containers to prevent lambs drowning.

Signs of lambing

It pays to know what to look for in a ewe's behaviour as lambing draws near.

If you have a good idea of when your ewes were served, keeping regular checks on them from just before the due dates is good practice. Four-hourly checks are good, if you can manage them. Chances are, if this is your first lambing, you will be checking on them even more often,

Left: Checking on expectant ewes.

Rachel Graham

Above and above right: Ewes showing early signs of labour.

Right: Inside the ewe: a full-term lamb.

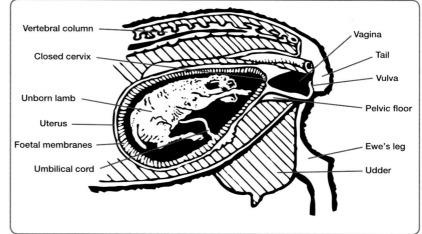

Vertebral column

Closed cervix

Unborn lamb

Uterus

Foetal membranes

Umbilical cord

Vagina

Tail

Vulva

Pelvic floor

Ewe's leg

Udder

eager to catch the action. When you approach the lambing paddock or shed, do so quietly, so you can observe your flock's normal, relaxed behaviour before they spot you, and then make every effort to avoid disturbing or startling them.

Over the weeks, you will have noticed your ewes putting on weight, particularly growing wider, and you may have spotted the movement of the unborn lambs inside them from time to time. The udder will have begun to swell, too, 'bagging up' and starting to fill with milk. The teats may look more prominent a day or two before lambing and the vulva will start to swell and darken.

The first sign that something may be happening is the ewe attempting to distance herself from the rest of the flock – which is one reason why expectant ewes should have plenty of space when kept in groups. If she's outside, she will probably head for a far corner of the field with a hedge or wall for shelter; inside she will do much the same, seeking out a quieter part of the building. She will be reluctant to eat and is likely to be restless, getting up and lying down repeatedly as if she is uncomfortable (because she is, due to the contractions). She may start pawing the ground with her front feet, as if attempting to scrape the grass or straw backwards to make a nest. When lying down, she may be seen 'stargazing' – twisting her head around and skywards (above left). She will show signs of straining, her breathing will quicken and she may grind her teeth.

Dystocia: coping with birth complications

Most births will be straightforward, and will often occur before you arrive. However, there may be times when a ewe will suffer from some form of dystocia – difficulty giving birth – and will need assistance. Knowing when to intervene and when to leave well alone comes with experience.

It's generally accepted that you should intervene when:

■ The water bag appeared or ruptured 30 minutes ago, but the labour has not progressed;
■ The ewe has been restless and straining for more than an hour;
■ Just a head is visible – no limbs; or
■ A single leg or a tail are the only things visible.

THE BIRTHING SEQUENCE

A thick string of mucus will be the first thing to appear from the vulva. As the contractions continue and get closer together, the cervix dilates to full capacity and the amniotic sac – the water bag protecting the head – will appear. It normally ruptures soon afterwards, at which stage the lamb's nose and feet should be visible, as long as it's lying in a normal position (see illustration page 112). Contractions inside the ewe should then start to push out the lamb. Sometimes, when the ewe is lying down, the sac doesn't break until the lamb is out and the ewe stands up. Lambs are often born within 30 minutes of the amniotic sac appearing, but this can vary.

Some ewes will be more comfortable delivering while lying down, while others choose to stand. Try and make sure that lambs have soft bedding to land on, just in case.

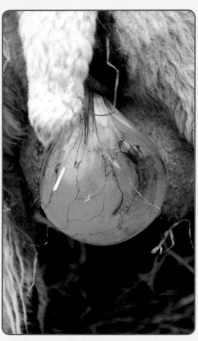

1 The water bag emerges.

2 The feet and nose begin to appear.

3 The lamb appears head first, with its nose and mouth still partly covered by the amniotic sac.

4 As the slippery lamb emerges, gravity helps it on its way.

All pics Rachel Graham

5 Safely down to earth, the lamb's face is now clear of the amniotic sac.

6 The ewe begins licking her newborn, starting the bonding process.

7 As the cleaning-up process continues, the placenta starts to appear.

An examination should only ever be carried out when the ewe's cervix is fully-dilated, otherwise you will not only cause her unnecessary pain, but you also run the risk of doing serious internal damage – and even death. If you do intervene, hygiene is vitally important. No one will expect your lambing shed to be as clean as a hospital operating theatre, but every effort should be made to prevent infection.

If the fleece around the vulva has not already been trimmed and is contaminated with faeces, use dagging shears to tidy her up and reduce the risk of infection. If possible, wash the area with a gentle disinfectant solution. The last thing you want to do is introduce bacteria into the ewe, which can lead to metritis, an inflammation of the uterus, which can prove fatal. Take off any rings or watches and make sure nails are clean, short and not jagged. Disposable gloves are preferable to bare hands, but hands should still be washed in case the gloves tear, exposing your skin. Gloves are much more hygienic than bare hands, and also give protection against zoonotic infections (see chapter 6) which are diseases that can pass from animals to humans and vice-versa. Although short plastic gloves are fine for shallow penetration of the vagina, shoulder-length ones are preferable, because some lambings will require you put your arm in up to the elbow. An obstetric lubricant should be used liberally, whether you are using gloves or not. Make sure to wash hands well after any examination and certainly before contact with any other ewes or lambs.

Carrying out an internal examination

After lubricating your (preferably gloved) hands, put your fingertips together, making a kind of cone shape, to make entering the vagina easier for both you and the ewe. Should you need to change hands, remember to wash and lubricate all over again. If the ewe is fully dilated, you should feel the

Below: Learning to deliver, using a dead lamb and a 'dummy ewe' made from a cardboard box.

nose of the lamb. The next task is working out the position of the legs.

The majority of lambs are born in the correct position, which allows an easy exit. This means coming head-first with the two legs forward, either side of the nose (see illustration on page 112). If it looks as if the labour is progressing normally, let the ewe get on with it and give her some space, but keep a watch just in case she needs a helping hand.

Ewes need assistance for a variety of reasons. In the majority of cases, it's either because the lamb's limbs are in the wrong position, or because it's too big to come out naturally – often the case with young ewes giving birth for the first time, over-fat ewes, or when lambs have been sired by strong, muscular breeds, which produce larger heads and shoulders. In extreme cases, a vet may need to carry out a caesarean section.

Always call your vet if you don't feel sufficiently confident to deal with a lambing dilemma.

Have someone hold the head and shoulders of the ewe if possible. Sometimes, making the ewe stand, or propping up her back end, makes some space for repositioning the lamb, as gravity plays in your favour. If the ewe has been in distress for a while and you have tried for 15 minutes or so to correct the problem with no success, you should call your vet.

Knowing how to identify the front and hind legs entirely by touch is essential. When assisting a ewe that you know or suspect is having twins (or more), the problem of untangling limbs can be more difficult, so you must be certain of which ones belong to which lamb before attempting to rearrange anything.

Study the anatomy of a newborn lamb so you know what you need to be feeling for. Close your eyes and feel the differences between the front legs and the back ones. You'll notice that the straighter forelimbs feel completely different to the hind ones, which have distinctive hocks.

Feel your way up to the top of a foreleg from the hoof and continue moving your hand up and you will feel the

Above: A lambing student at Humble by Nature carries out an internal examination.

sharpness of the shoulder blade. Move your hand across the shoulders – without losing contact with the lamb – and find the opposite shoulder blade and then work your way down the other front leg. When you identify a hind leg, work your way up to find the pelvis and you should be able to locate the tail just behind. Now feel your way across the pelvis and down the other hind leg to again satisfy yourself that both legs belong to the same lamb.

A FORELEG

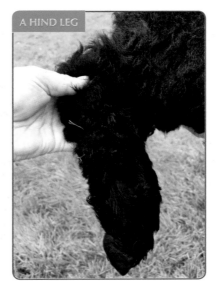

A HIND LEG

DEAD OR ALIVE?

How do you tell if the lamb you are trying to deliver is alive? Stick your finger in its mouth and, if it's a healthy lamb, it should try to suck. If nothing happens the lamb is either dead or very weak indeed.

ASSISTING A BIRTH

1 After checking the lamb is in the normal position, the front legs are held.

2 The legs and head are eased out.

3 The lamb is delivered.

4 The ewe is encouraged to bond with her lamb.

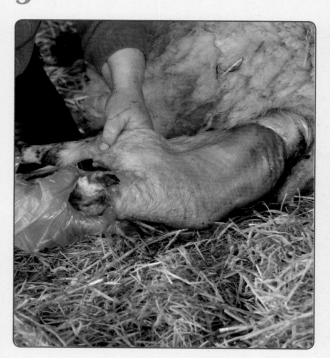

All pics Rachel Graham

Malpresentations

In each of the cases illustrated here, the lamb will normally need to be repositioned manually. Most of the time, legs that are tangled or wrongly-placed can be corrected with careful manipulation, but some complications are more difficult than others. There will be occasions when the head has already emerged, but the entire lamb needs to be gently pushed back inside the ewe in order for repositioning to take place.

In any assisted delivery, cleanliness of hands, gloves and any equipment used is essential. An injection of long-acting antibiotic should be given afterwards as a precaution against infection.

Front leg (or legs) back

Probably the most common lambing problem to occur is a forward-facing lamb stuck because either one or both front legs is tucked back. Inserting your lubricated hand, feel your way down the shoulder to locate each leg in turn. Cup each foot in your hand, to prevent the hoof damaging the wall of the uterus, and straighten it out. When helping to deliver a lamb, work with the ewe's contractions – not against them. Pull it out in a downward curve, towards the feet of the ewe.

Backward-facing lamb (posterior position)

Fairly common in multiple births, the lamb's hind feet will be felt first. Don't attempt to get it facing the opposite way, for a normal presentation – pull it out by the hind legs. On rare occasions, a backward-facing lamb will also be lying upside down – with its belly facing the ewe's spine. Pulling in a downward curve in this situation could damage the lamb's spine, so pull it straight out instead.

A backward-facing lamb needs to be delivered quickly, as there is a risk that, if the umbilical cord breaks while the lamb is still inside the womb, it will start to breathe in fluids.

Breech presentation (abnormal posterior position)

A more complicated variation of the previous situation – particularly if the lamb is a large one. A backward-facing lamb sometimes has its hind legs folded underneath itself – pointing forwards, towards the head of the ewe, instead of being stretched out towards her back end. In this scenario, cup each hind foot in your hand and ease it towards you, taking it into the birth canal, and then deliver as you would in the case of a 'normal' backward-facing lamb.

Head turned back

A lamb that has one or both front legs forward, but the head twisted backwards, can be tricky to deliver. You may be able to cup the head in your hand and ease it round, but it may slip back. It may be necessary to secure a lambing snare behind the ears and through the mouth (NOT around the lamb's neck), and then lambing ropes around the legs. When the lamb is pulled forward into the pelvis, it's unlikely to slip back and it should be able to be delivered by pulling gently and steadily on the snare and ropes.

Delivering twins

As long as you know how to distinguish between forelegs and hind legs, and how to work out which legs belong to which lamb (see page 110), delivering twins should not present any great difficulty. Often, once the first has been repositioned and delivered, the second will arrive normally – although you might as well finish the job and deliver both. Note that twins can be found lying in different positions – either one pointing forward and one pointing back, or both pointing forward. Remember this when you are feeling for limbs, as it's easy to get confused.

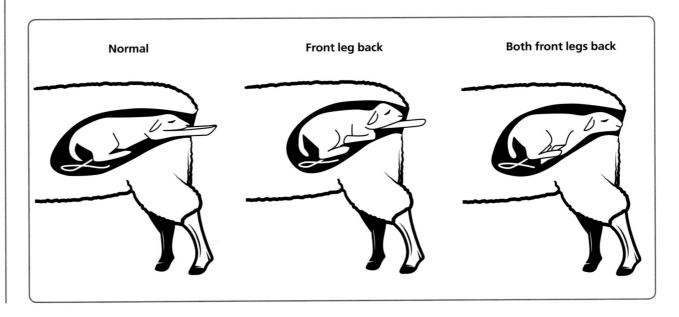

Normal   **Front leg back**   **Both front legs back**

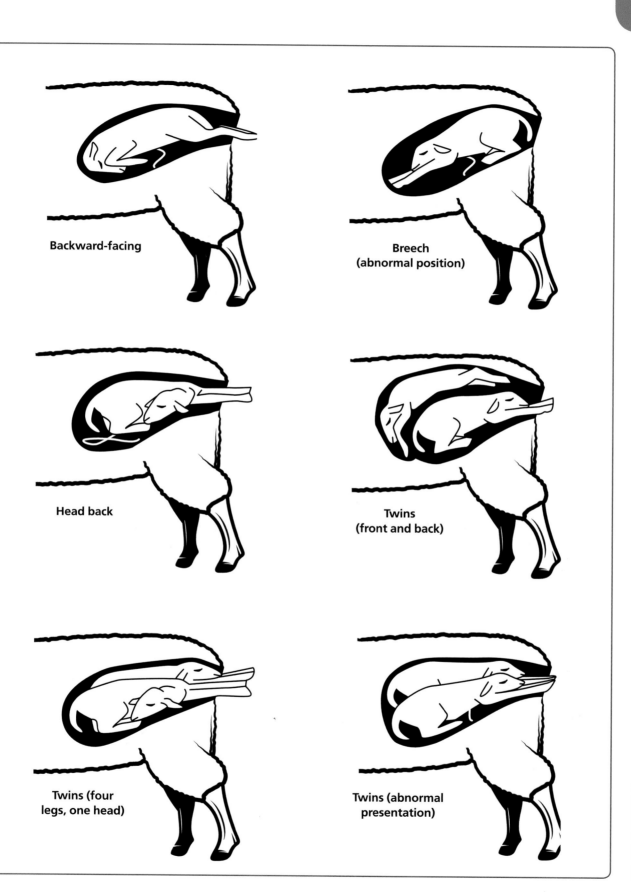

Backward-facing

Breech
(abnormal position)

Head back

Twins
(front and back)

Twins (four
legs, one head)

Twins (abnormal
presentation)

After delivery

Even with a normal birth, there are some occasions when the amniotic sac won't rupture, in which case you may have to break it open to prevent the lamb from drowning in fluid.

Once delivered, check that the nose and mouth are clear of afterbirth and mucus and that it's breathing well. Shepherds often gently insert a piece of straw up the nostrils to make the lamb sneeze, cough, or shake its head, clearing any blockage. A vigorous, stimulating rub is sometimes all it takes to perk up weary lambs, but there are occasions when more help will be needed. Oral preparations such as Lamb Kick Start are useful to keep close to hand in case weak lambs need a boost of glucose to get them up and suckling.

Swinging a lamb by its back legs may look a bit odd – and slightly dangerous, too – but it can be an effective way of clearing the airways of mucus. The lamb will still be wet, and therefore slippery, so make sure you hold it tight!

Move the lamb around to the ewe's head so that she can start cleaning it and bonding with it. Allow this bonding process to happen before carrying out any procedures like treating the navel.

Premature births and orphan lambs

Often linked with multiple births or poor ewe nutrition during pregnancy, but also associated with infectious diseases, e.g. enzootic abortion. Lambs are small, weak,

Below: Swinging a lamb after a long labour.

Above: Clearing mucus from the nose and mouth.

frail and prone to hypothermia – partly because of their thinner, poorly-developed coats. Their weakness may mean they can't get up and suckle – and even if they are ready to, they may still suffer because the ewe will not have produced colostrum.

Breathing problems may occur, too, because the lungs won't be fully developed and so are not able to expand and function properly when the first breath needs to be taken. When still inside the ewe, oxygen is supplied via the placenta, with the lungs taking over when the lamb is born. If the time between birth and taking the first breath is too long, the lamb can die.

Care of the umbilical cord

This is a very important procedure to avoid infection entering the body and causing problems like joint ill (see page 121). Dip the cord in tincture of iodine or use antibacterial spray to protect it. Some sheep keepers will simply spray the navel, but dipping is more effective to ensure complete coverage. The cord will begin to dry quickly and will fall off after a few weeks. The ewe will occasionally nibble at the end of the cord while cleaning her lamb. It's unusual to have an excessively long umbilical cord, but if it does appear longer than normal and might cause problems, clip it shorter with sterile scissors and dip immediately. Plastic navel clamps, which clip tight at the base of the cord, can also be used. Some people prefer to use these instead of pungent iodine because some ewes can be confused if their lamb smells strange and may not recognise it as their own.

Caring for newborns

Ensuring that lambs receive an adequate intake of colostrum – the first milk produced, which is rich in antibodies and essential nutrients – is vital as soon as possible after birth. In ruminants, there is no transfer of antibodies across the placenta, which means that lambs are born without protection against disease and so depend on colostrum for passive immunity.

Whether it comes in its natural form, straight from the

Above: Allow the ewe time to bond with her lamb.

ewe, or as a bought-in powder feed, which is mixed and bottle-fed, colostrum is a must in order to ensure that the lamb gets the best possible start in life. As well as being a source of protection against infection, colostrum helps protect the lamb from hypothermia. Lambs are born with limited supplies of what is known as 'brown fat' – a kind of fuel that they use to regulate and maintain body temperature until they get their first feed.

Colostrum must be consumed during the first few hours of life to have a beneficial effect because, afterwards, the body is unable to absorb the antibodies as effectively. A lamb should be fed a minimum of 10% of its body weight in colostrum in the first 24 hours of life – with at least 50% (approximately 150g to 180g) being given in the first six hours.

Recommended colostrum intakes
Large single lamb (c.5kg): 250 ml, 3 times a day
Medium lamb (c.4kg): 200 ml, 3 times a day
Small lamb (c.3kg): 100 ml, 4 times a day

While feeding colostrum to the lamb is important, it is just as vital to discover why the lamb could not feed in the first place. The lamb may have simply been rejected by the ewe, but it may not have suckled because the udder was inflamed with mastitis. Care of the ewe, as well as the newborn lamb, is equally important.

Dead lambs

If a ewe has problems delivering and has a dead lamb in the uterus, it may be difficult to get out. The ewe will stop making an effort to push, and the uterus is likely to be dry; plenty of lubrication will be needed to avoid both damaging the lining and hurting the ewe. Depending on how long the lamb has been dead, the ewe may need veterinary attention. A course of antibiotics may be prescribed, along with an injection of oxytocin, to induce contractions of the uterus and expel the placenta and any other debris, such as parts of the decomposing lamb.

Potential ewe problems at lambing

Ringwomb
This is a condition in which the cervix does not dilate enough to allow lambing. Even though the lamb is ready to be born, examination will reveal just a small opening, about 3 to 5cm in diameter. If several hours have passed since the first signs of mucus, your vet should be called. Muscle relaxants are sometimes used but often prove ineffective and the only real option available is a caesarean section. Other conditions can appear similar to ringwomb, such as a tight cervix due to the ewe being overweight – another reason to carefully monitor feeding during pregnancy.

Uterine inertia
Uterine inertia – a condition in which the muscles in the uterus fail to contract – is less common in sheep than in some other species. There are two main types of inertia: either the uterus refuses to contract at all, despite the cervix being dilated; or, after a long labour when the ewe has been having contractions and actively been trying to deliver, the muscles tire and appear to stop working. In both cases, the lambs need to be delivered as soon as possible, using plenty of lubrication. If you can't do it yourself, call the vet or take the ewe to the surgery.

Uterine torsion
This is a twist of the uterus, which prevents the lambs being expelled. The first sign is that a ewe has entered the first stage of labour, but has not been able to deliver. Again, a rare occurrence, this condition effectively blocks the birth canal; if you carry out an internal examination, you will feel the twist. It can't be corrected by internal manipulation – and this should never be attempted. Some cases have been remedied by physically rolling the ewe in the opposite direction to the twist, but this can stress the ewe and, with the prime aim being to get the lambs out as soon as possible, the vet should be called as a matter of urgency. A caesarean section may have to be carried out.

Retained placenta
Placentas normally come away fairly soon after a birth, but it can take 12 hours or more to be shed. Retention can be caused by various factors, including abortion, stillbirths and infection. Nutritional deficiencies (e.g. selenium, calcium, vitamin A or E) may also have an effect. Part of the placenta will normally be seen hanging out of the vagina and the ewe may appear depressed and lacking in appetite. Behaviour is a key indicator of something being wrong, but you may notice a bad smell as the placenta starts to decompose.

NEVER pull a placenta because it may still be attached to the uterus and cause serious damage and risk infection. Your vet may recommend an injection of prostaglandin or oxytocin, to stimulate expulsion, along with antibiotics.

Prolapse

Two types of prolapse can occur – of the vagina and of the uterus. Vaginal prolapse most commonly happens in the final three weeks before lambing. Although it's more likely in older, overweight ewes, it can affect ewes of any age. The first sight is normally when the ewe is lying down; the prolapse may start by making 'guest appearances' – slipping in and out at various times, but gradually spending more time visible. The condition often puts so much pressure on the urethra – the connection between the bladder and the vagina – that the ewe can't urinate, and this worsens the condition and makes the ewe even more uncomfortable and causes more straining.

The sight of a bright red, fleshy prolapse – which can range from the size of an orange to that of a small football – can be alarming, but the condition can be remedied with the right care. If in doubt, always consult your vet. There are various ways of treating the condition, and the first step is getting it back into place – which is not always an easy job. It's sometimes easier to do this if the ewe's back end can be elevated so that gravity helps relieve the pressure caused by the abdomen.

Hygiene is important – particularly if the ewe has been lying on the floor and the prolapse has come into contact with the ground – so carefully wash the prolapse with warm

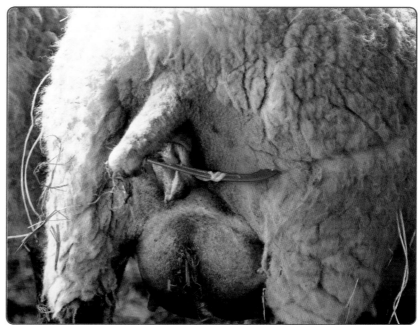

Rachel Graham

Above: A prolapse spoon in place.

water and a gentle skin cleanser or surgical scrub solution, use plenty of lubrication, and gently ease the mass back in. When the prolapse moves back into position, the ewe may urinate in relief. A course of antibiotics should be given to guard against infection.

Once the prolapse is back in place, it needs to be secured. The most common methods include inserting a plastic retainer or 'spoon', tying string or wool across the vulva, attaching a special harness to exert external pressure on the perineum, or suturing the lips of the vulva and monitoring closely as lambing time draws near. These makeshift arrangements need to be removed once the lambing process is underway. Ewes will, more often than not, lamb successfully after a prolapse, but supervision is recommended. These ewes should be considered 'high maintenance' and a decision should be taken as to whether or not to cull after weaning.

Prolapsed uterus normally happens at lambing or pretty soon afterwards and, in quite a dramatic fashion, the entire uterus is visible hanging from the back end of the ewe. Call a vet, because this is a much more complex problem to rectify. Try and keep the prolapse clean and the ewe contained in a small area until the vet arrives. The likelihood is that the ewe will need to be culled.

Mastitis

This is an uncomfortable inflammation of the udder, which can lead to death if left untreated. The cause is bacteria finding its way into the teat canal and entering the udder. The majority of cases are caused by bacteria, which are always present on the skin surface or carried in the mouth. It's common for lambs to pick up the bacteria from the ewes

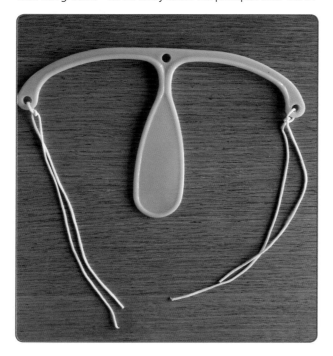

may be able to feed a single lamb, they would not be able to cope with twins or more.

Treatment for mastitis, if the condition is caught sufficiently early, is by antibiotic injection, and your vet may also prescribe anti-inflammatory drugs.

Prevention is always easier than cure, so observe good hygiene at lambing time. For instance, always ensure you have clean hands when checking that a newly lambed ewe has milk, so avoiding transferring bacteria to the teats. Damage to the ends of the teats can be caused by lambs' teeth or by lambs sucking too hard or butting the udder in an over-enthusiastic way – often a sign that the ewe is short of milk.

Cold and windy conditions are also thought to increase the likelihood of mastitis, so providing shelter is a worthwhile precaution.

and transfer to the teats during suckling. Another cause can be environmental bacteria, such as E. coli, which is widespread in pens that are damp, dirty, overstocked with sheep, or which have insufficient bedding. In some rare cases, mastitis can be caused by the Maedi Visna virus or leptospirosis (see Chapter 6).

Most mastitis occurs either really soon after lambing or three to six weeks later, when milk production is at its peak. Any ewes that fall ill after lambing, have poor appetite and look depressed, or whose lambs appear hungry and ill-nourished, should always be checked for mastitis.

Mild cases may see the ewe off her food and reluctant to get up. She may appear to be suffering from lameness in one of the hind legs, as she tries to swing a leg away from the painful udder. You may notice her shaking her hungry lambs off as they attempt to suckle and the milk, when expressed, may be thin and watery or could contain blood or clots. On examination, the affected half of the udder will be swollen hot and the ewe will show signs of pain when it is touched.

In the most severe cases of mastitis, the blood supply to the affected half of the udder is disrupted. The udder becomes increasingly swollen and hard, and changes in colour to dark red, purple, or even black. At first, the udder is likely to feel hot, but as the tissue becomes gangrenous it will turn cold. If the ewe doesn't die, the affected side of the udder will drop off in time. Culling is the only real option, as the ewe is unlikely to be able to rear lambs again. In addition, the exposed wounds on the udder will take a considerable time to heal and can become infected and will attract flies in summer.

Some less serious cases are only discovered when ewes are checked over prior to mating. Udder checks should be a routine part of the pre-mating MOT, and it may be that lumps of scar tissue will be felt in the udder or in a teat. Such ewes should be moved for culling because, although some

Problems with newborn lambs

Rejection, orphans and fostering

Not all lambs will be able to feed naturally and sometimes nature needs a helping hand. Some ewes will simply not accept their lambs willingly, while others may accept one twin but not the other.

Sometimes it's simply a case of holding the ewe steady and letting the lamb suckle for a few minutes so the ewe gets used to the experience and bonds with the lamb. This can be done with the ewe standing or tipped on her bottom. Another option is to put her in a small, purpose-built pen – often called a lamb adopter – and maybe restrain her head using a purpose-made yoke, which operates rather like medieval stocks. Ewes that reject their lambs can be extremely aggressive towards them and will headbutt them away repeatedly, so restraint is often essential. This isn't a foolproof method, and sometimes the lamb has to be artificially reared. The ewe is normally kept restrained for 48 hours, but the lambs will have to be monitored to ensure they are actually feeding. Make sure that, when she is released, she has actually accepted the lamb and doesn't attack it.

Rejected, orphaned lambs, or 'spare' lambs from multiple births can often be fostered on to ewes that have either lost their own lambs or have given birth to a single lamb but have sufficient milk for two.

Fostering works well with strong, healthy lambs, which are likely to be persistent when looking for a feed. Weak or sickly

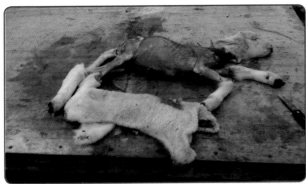

Angus John Mackenzie (both)

Above: (Top) A dead lamb's skin is removed and (bottom) fitted onto an orphan.

lambs are unlikely to be good candidates for fostering and are at more risk of being abused and injured by the ewe. Lambs that have been bottle-fed often find it difficult to adjust to natural feeding, while those that have been fed by stomach tube (see page 119) and have not developed a sucking habit will fare better.

One tried-and-tested way of conning the ewe into accepting a strange lamb is to make it smell just like her own. Covering the new lamb in birth fluids and the placenta from the foster mother is one technique. If she has one lamb, the 'bin method' may work. This involves putting both her own lamb and the new one into a clean dustbin with the placenta for an hour or more so that they both begin to smell alike. Some farmers cut the placenta in two and tie a piece around each lamb's neck.

If the ewe has given birth to a dead lamb, another option is to skin it and make a sleeveless 'coat' for the lamb that needs adopting, remembering to use any available birth fluids or the placenta to rub over the lamb's head, legs, anus and tail.

Mis-mothering and abandonment

Ewes close to giving birth sometimes get confused at the sight and smell of another's newborn lamb and try to 'steal' it as their own. This can happen when a ewe is giving birth to the second of twins and the firstborn wanders off. The danger is that a) the bond with the birth mother can be broken and the lamb may not be accepted back; and b) the ewe that is still to lamb may reject her own lambs in favour of the wandering one.

Some ewes – like humans and many other mammals – are simply not cut out for motherhood. Sometimes it can be an inherited trait, but a reluctance to bond with a lamb can be due to ewes being stressed or disturbed during or after giving birth. Young or inexperienced ewes are more likely to abandon their lambs, but older ones that are simply feeling hungry may also go off to look for food and not return.

Too much human intervention can unsettle a ewe, so try and avoid crowding her while she does her job. Unless the lamb is at risk because the sac has not broken, allow the ewe to clean off her newborn, as this is a vitally important part of the bonding process.

If a ewe is tired after giving birth to twins or more, the final lamb may end up being neglected.

Hand-rearing

Ideally, colostrum from the lamb's mother should be used, or some taken from another ewe in the same flock. Cow colostrum (often used in powdered colostrum supplements) is also a good substitute, as is goat colostrum. Dried colostrum can be bought in a wide range of quantities, from single sachets, which mix with water to make one feed, to large tubs suitable for providing hundreds of meals.

Lambs that are being hand-reared should be fed four to five times in the first 24 hours, with at least two of those feeds being colostrum and the remainder specialist milk replacer powder. Milk replacer is designed to be a complete diet, providing the lamb with all the energy and nutrients it needs. From day two, milk replacer can become the only feed, and should be given three times a day (e.g. 6am, 2pm, 10pm). If at all possible, spread the feed between a greater number of sessions (e.g. 6am, 12pm, 6pm, 12am), feeding smaller quantities to make digestion easier. In the first week of life, a 5kg lamb should be taking 1 litre a day, and thereafter 1.5 litres a day, with feeds gradually increasing in size and becoming more spread out. Adjust the quantity if lambs are particularly small, working on the basis of 50ml per kg of weight per feed.

Bottle-feeding

Most lambs take easily to the bottle and, even though it's a

Below: Feeding orphan lambs.

Left: Lamb drinking from an ad-lib feeder.

Right: A lamb being fed by stomach tube.

Rachel Graham

Claire John

time-consuming and – if you're buying in milk replacer – an expensive business, it can also be incredibly rewarding. Bottle-fed, 'pet' or 'cade' lambs can be a useful addition to the flock, too. They can be a bit of a nuisance, as they attach to the human who feeds them incredibly quickly. On the other hand, as they grow up trusting you and will come to you when called, they can be useful allies when trying to gather your sheep.

With all methods of artificial feeding, pay close attention to hygiene at all times to avoid infection.

When mixing milk replacer, follow the instructions on the pack carefully and, if you're using hot water, make sure the milk is cool enough before feeding. Test it on your wrist, just as you would with milk for a human baby.

Lamb feeding bottles are available from agricultural merchants, but a standard baby's bottle is just as practical, if slightly smaller. Teats come in different shapes, sizes and materials. When making a hole in the end, make sure it's not too big, otherwise the flow of milk will be too fast for the lamb to cope with. Always hold the bottle as upright as possible, to avoid the lamb sucking in air instead of milk.

Inhalation pneumonia
Careless bottle-feeding of weak lambs – lambs that should have been tube-fed (see below) – is the most common reason for this respiratory problem. Lambs with a poor sucking reflex are often the victims, as milk accidentally dribbles into the windpipe and leads to an infection in the lungs.

Sometimes the first sign of this condition is finding the lamb dead. If found alive, the lamb will be weak and will be having trouble breathing, and you may be able to hear a crackling sound in the chest. Antibiotics are needed to treat the condition, so consult your vet immediately.

Additional feed
Fresh water should be available at all times and creep feed should be offered when the lamb is a week old. Roughage such as hay should not be offered, as it can reduce consumption of milk replacer and creep feed, having an impact on growth and potential weaning date.

Before you consider weaning, all artificially raised lambs should be at least 35 days old and should be eating at least 250g of solid feed per day. They should also, ideally, weigh 2.5 times their birth weight. Although many people like to reduce milk replacer gradually, studies have shown that abrupt weaning is a better way of reducing the chance of digestive upsets.

Tubing
Always look out for signs of hungry lambs. If a lamb has not fed within the first two or three hours of life – for whatever reason – it may be necessary to feed colostrum by stomach tube. Note, however, that you should NEVER use a stomach tube on a very weak or unconscious lamb, as there is a real risk of the tube entering the windpipe.

Getting the tube inserted in the correct position is the biggest concern for newcomers to shepherding. Have the colostrum ready and warmed to body temperature. Holding the lamb, slide the tube into the side of the mouth, but without forcing, until 50–75mm (2–3in) is still visible, or until resistance is felt. The lamb should start to swallow, but should it show signs of distress, remove the tube and try again. Once the tube is in place, attach the syringe and slowly start to give the colostrum over 25 seconds before removing the tube. Lambs usually respond quickly to being tube-fed.

Hypothermic lambs
A lamb's normal body temperature is 39–40°C (102–104°F), and anything below 37°C (99°F) is considered to be hypothermic and in need of being warmed quickly. Newborn lambs lose body heat very quickly if they are not dried quickly by their mothers, and also if there is a delay in suckling. Always dry the lamb first before attempting to warm it. Avoid using an infrared lamp because the heat can't be controlled and the lamb can become *hyper*thermic (overheated) and also there is a risk of burning the skin if the lamp is too close. Specialist lamb-warming boxes are available, but it's commonplace to see lambs taken into warm rooms in the house until they warm up sufficiently. Return the lamb to its mother as soon as possible, to minimise the risk of rejection.

How to tackle hypothermia

Temperature and age	Action
37–39°C (99–102°F) – any age	Dry the lamb's fleece. Administer feed by stomach tube. Return to the mother or house with other lambs. Check temperature at intervals.
Lower than 37°C – birth to 5hrs old	Dry the lamb's fleece. Warm until temperature reaches 37°C. Feed by stomach tube. Return to the ewe or place in your lambing 'sick bay' for monitoring.
Below 37°C – more than 5hrs old and able to support its own head	Dry the lamb's fleece. Feed by stomach tube. Warm until it reaches 37°C. Feed again by stomach tube. Return to the ewe or place in your lambing 'sick bay' for monitoring.
Below 37°C – more than 5hrs old and unable to hold up its head	Dry the lamb's fleece. Give an intraperitoneal injection of glucose (ask the vet to do so if you're not confident). Warm until it reaches 37°C. Feed by stomach tube. Return to the ewe or place in your lambing 'sick bay' for monitoring.

Scouring (diarrhoea)

Milk replacer can cause lambs to scour because it often contains a laxative ingredient to encourage them to pass their first faeces – which is known as meconium. However, scouring can also be due to bacterial infection, a virus, or internal parasites. It's more commonly seen in flocks that are housed for lambing, as hygiene standards can be more difficult to uphold.

In simple terms, diarrhoea is a condition in which the body's faeces – normally solid waste – becomes more liquid than usual and is excreted from the bowels more often. It can vary in colour, consistency and severity.

Some lambs will appear lethargic, go off their food, lose weight and have a sunken-eyed look. Dehydration is the key problem, as too much liquid is leaving the body and not being replaced, as the lambs tend to be less active and drink less. Dehydrated lambs can go downhill very quickly indeed, and some never recover.

Bacterial infection is one of the most likely causes in lambs and if they don't suckle sufficiently well, they are at more risk. At birth, lambs have little defence against disease and bacteria like *E. coli*, *salmonella* and *Clostridium perfringens* can pose a major risk. Colostrum, the ewe's first milk, contains antibodies that give initial protection, which is why it is so important that lambs suckle as soon as possible. Unfortunately, this defence shield doesn't last forever, and

also, if there are very large levels of bacteria in the lambing shed, even the maternal protection can be defeated. Other factors, such as hypothermia and stress, can make matters worse, lowering resistance even further.

Bacterial infections will respond to antibiotic treatment, but viral infections won't. The bad news about viral diarrhoea is that there is no treatment, so all you can realistically do is provide supportive care such as warmth and electrolytes in drinking water. An electrolytes solution such as Liquid Life Aid – which contains ingredients such as glucose, salt and potassium – should be added to drinking water to help combat dehydration and restore the balance of fluids and minerals in the body.

The most serious type of scouring – lamb dysentery – is caused by the bacterium *Clostridium perfringens* and can bring swift death within a matter of hours of the first signs (bloody diarrhoea) being spotted. Fortunately, ewe vaccines are available (see Chapter 6).

Parasites such as *Cryptosporidium* and *Giardia* spp. can also be the cause of significant scouring, but are often overlooked as the culprits. Transmission can be due to ingestion of faeces, or contaminated soil or water. Some species can infect humans and vice versa so good personal hygiene after handling lambs is essential.

Cryptosporidiosis can affect lambs from one week old and some become very ill and can die through dehydration if fresh water is not readily available. There is no treatment. Giardiasis is often present without clinical symptoms. Lambs often lose weight and are slow to reach slaughter weight. The wormer, fenbendazole, has been shown to be highly effective. In the case of both *Cryptosporidium* and *Giardia*, oocysts shed by lambs can remain active for over a year, ready to infect next year's new arrivals.

Watery mouth

Also known by other names including rattle belly, slavers and slavery mouth, this condition is normally seen in lambs from 12 to 72 hours after birth. It's another problem that leads back to poor hygiene; the lamb ingests large amounts of bacteria – often *E. coli* – which cause blood poisoning. The lamb becomes lethargic, depressed and shows classic signs of a wet mouth caused by drooling saliva from the sides of the mouth. It stops suckling and quickly appears bloated.

Treatment is by antibiotics and feeding with an

electrolytes and glucose solution, given by stomach tube. On commercial farms an oral antibiotic preparation is often given within the first 15 minutes after birth as an additional protection against bacteria.

Working to prevent scouring and associated problems is easier than trying to get rid of them. Good hygiene practices in lambing sheds will help prevent infection building up. Clean and disinfect pens regularly and isolate lambs that show signs of scouring.

Joint problems

Sometimes lambs are born with leg problems, more often than not the front ones. They may be seen walking on their knuckles because they can't straighten the leg joints. In some cases, applying a padded splint or a firm bandage may help, but if the deformity is too severe, it may have to be destroyed. If you do decide to persevere and try and treat the problem, you may have to artificially feed the lamb because it will be less able to suckle. Joint defects in the odd lamb are not unusual and are normally regarded as 'just one of those things'. However, if numerous cases occur, it could be because your flock has been affected by Border virus, so consult your vet and ask for tests to be carried out.

Joint ill

Sometimes known as navel ill, pyosepticaemia or infectious polyarthritis, joint ill is a far more common thing to see. It's a bacterial infection, which causes pain, swelling and lameness, and one that can cause permanent joint damage if not treated quickly. Lambs quickly look depressed and lethargic and are reluctant to feed. As touched on earlier, treating the navels of newborns with a tincture of iodine solution or an anti-bacterial spray is important to prevent bacteria using the umbilical cord as a means of getting into the body. Good standards of hygiene in the lambing area are also essential to reduce risks.

If spotted at an early stage, treatment with antibiotics is effective, but the longer the infection is in the joints, the less chance the lamb has of survival.

Clostridial vaccines given to ewes (see Chapter 6) can prevent a variety of diseases, including lamb dysentery, a fatal disease caused by *Clostridium perfringens,* which often causes a bloodstained scour quickly followed by death.

Bloat

Bloat is a build-up of gas in the rumen and is more common in artificially fed lambs up to a month old than in others. The gas is a normal by-product of the digestion process and is normally lost by belching (sometimes described as eructation). When something prevents the gas from being expelled, the abdomen will begin to swell and the lamb may bleat in pain. It may die within 15 minutes of the condition becoming established, so the first sign may be a dead lamb.

Treatment methods vary. A stomach tube gently fed into the abdomen may allow gas to escape, but some other home remedies include feeding a heaped teaspoon of soap

Susie Kelly

Above: A lamb killed by bloat.

powder, or a small (2–3cm) block of margarine, to encourage the lamb to belch. A vet may also try a surgical incision to release the gas.

In the case of bloat caused by froth in the abdomen rather than just gas, a stomach tube can be used to feed in vegetable oil – between 10 and 12ml per kg of body weight.

In chronic cases of bloat, vets will often give antibiotics to control the bacterial fermentation in the rumen.

Bloat is best prevented by avoiding feeding large quantities of milk to lambs – particularly if the milk is warm. This is why it's recommended to divide the daily ration into several small feeds, to make digestion easier. Longer gaps between feeds also mean lambs may become hungry and start nibbling at rubbish in the pen, from baler twine to bits of wool – objects which can cause a blockage. Dirty equipment (feeding bottles, teats, stomach tubing equipment) can introduce bacteria and make the condition worse.

Infectious keratoconjunctivitis (pink eye)

This is sometimes described as 'snow blindness' or 'heather blindness' because it's often associated with sheep on high ground in adverse weather conditions. However, it affects sheep reared both outdoors and in. The infection causes a painful inflammation of the conjunctiva, which can become so bad that either one or both eyes close. The cornea can also be infected and take on an opaque appearance. In the most serious cases, the cornea becomes ulcerated and the eye may rupture, resulting in permanent blindness.

Once in a flock, this infection is quickly passed from lamb to lamb, often through direct contact or from feeding together at troughs. Antibiotics are used to treat the condition – either by injection or eye cream, or sometimes both. Stopping the spread is the biggest challenge so separate affected lambs. There may be occasions when all lambs – even those not infected – will need to be treated as a precaution. Unfortunately, it can be a recurring condition.

Above: A lamb with entropion.

Rachel Graham

Entropion

This is a condition where the lower eyelid turns inwards to irritate the cornea, causing discomfort and weeping. It's often seen at birth or soon after and, if left untreated, the irritation can lead to ulceration and blindness. When the condition affects both eyes, the lamb's eyes may be completely or partly closed and it may have problems finding the udder and suckling.

It's an inherited condition, more prevalent in some breeds than others, and selective culling may be the best option if many lambs are affected. Various treatments are used, including carefully injecting a small amount of antibiotic into the eyelid. It's a tricky job requiring one person to hold the lamb steady while the other carries out the subcutaneous injection, and should only be carried out if you know exactly what you are doing. Otherwise, call your vet. The weight of the antibiotic injected (normally just 0.5ml) is enough to turn the eyelid back out, and it also protects against bacterial infection. A popular 'quick fix' remedy often used by farmers is to pinch the eyelid with the fingernails, so causing a swelling, which prevents the eyelid from rolling back in. Surgical clips called Michel clips are another option. These are inserted into the skin under the eyelid to weigh down the skin and pull the eyelid into the correct position.

Pulpy kidney

This is a gut infection caused by clostridial bacteria. Toxins in the bloodstream damage internal organs, including the heart and the brain. Lambs are often found dead, without there having been prior signs of illness, but live lambs may appear increasingly excitable. Vaccinating ewes for clostridial diseases is recommended.

Tetanus

Clostridial vaccines given to ewes can help prevent this condition, which is caused by a toxin in the body produced by the bacterium *Clostridium tetani*. Often the route of infection is via a wound. Symptoms appear 3 to 10 days after infection and can initially be stiffness of the limbs and/or muscle tremors. Convulsions and breathing problems follow before death occurs. Only the less-severe cases can be treated, using antibiotics and tetanus antiserum, but, in most cases, the lamb will need to be destroyed at the first signs of stiffness.

Umbilical hernia (gastroschisis)

Quite an alarming condition for the newcomer to breeding, this sees the lamb's intestines hanging outside the body because of a malformation of the navel. The unborn lamb has a natural gap in the body wall to allow the blood vessels of the umbilical core to pass through. Normally, this gap is very small and closes soon after the lamb is born, but occasionally the opening is too large and the muscles do not seal completely, allowing the intestines to slip through. Sometimes just a small amount protrudes, but the condition can be made worse as the ewe attempts to lick the lamb dry, pulling more through.

A sign that an umbilical hernia is likely can be a swelling of the navel, but you are more likely to spot trouble when the intestines have already broken through.

Treatment is tricky and is a job for the vet. What you should aim to do is to keep the intestines clean and protected until a vet can make an assessment. You will need to take the lamb straight to the surgery because, unless treated, the lamb will deteriorate very quickly. When the condition becomes apparent in human babies, the first step is to wrap the abdomen in plastic cling film to guard against infection. An alternative would be a clean towel. It's likely the vet would give the lamb a general anaesthetic and attempt to make the gap in the body wall bigger to allow the intestines to be replaced, and then suture it closed.

If you can't get veterinary attention immediately, you may have to consider killing the lamb humanely on site, because the condition will not improve untreated.

Cerebellar atrophy (Daft lamb disease)

Sometimes mistaken for swayback (see page 89), this is a genetic condition in which part of the lamb's brain has not developed properly. An affected lamb will carry its head high, mouth pointing up to the sky or to the side. Sometimes it may walk in circles before collapsing. There's no cure and the lamb should be destroyed. Culling the parents should be considered as a preventative method for the future.

Above: A lamb with an umbilical hernia.

Management of lambs

Castration and docking

Castration is carried out to allow ewe lambs and ram lambs to be run together without the risk of unwanted pregnancy, and also to avoid accidental breeding from stock that is not considered good enough.

Whatever method of castration is used, the procedure can only be carried out by a suitably trained, competent person aged 17 or over. Only a veterinary surgeon can castrate a lamb that is more than three months old.

Two main methods are used – elastrator pliers and tight rubber rings, and 'bloodless' castrators. The use of rubber rings for castration and tail docking is restricted by law to the first week of life. Both procedures are undoubtedly painful, as anaesthesia and analgesia are seldom used. Neither procedure should be carried out in the 24 hours after birth, as it may disrupt bonding with the ewe and discourage the lamb from sucking vital colostrum, therefore risking conditions such as watery mouth. You should always ensure that lambs are fit and healthy and have consumed sufficient colostrum before applying a band.

When castrating, the tiny but strong rubber ring is pushed on to the prongs of the elastrator and then the handles are pressed to stretch the band wide enough to fit over the scrotum. Care should be taken to ensure that the testes are descended, that there is no swelling of the scrotum, suggesting a scrotal hernia, and that the rudimentary teats (male lambs have teats as well) are not caught below the band. If the band is placed too high, the urethra – the tube connecting the bladder to the penis – can be damaged and the lamb won't be able to urinate and will die. The handles of the elastrator are relaxed and the band is then eased off the prongs. Eventually, the band will cut off the blood supply and the area below will drop off. Always remember – two teats above the ring and two testicles below it. If you manage to get the ring in the wrong place, take it off and start again. Use a small object like a pen or a teaspoon handle to stretch the ring first, to avoid cutting the lamb's skin.

Older lambs can be castrated at around four to six weeks of age using a bloodless castrator or emasculator – brands include the Burdizzo and the Richey Nipper. This tool is used to clamp around each spermatic cord in turn, crushing and rupturing them. As the blood supply to the testicles is halted, they eventually shrink, shrivel up, and deteriorate. The lamb needs to be suitably restrained throughout the procedure and the instrument must NEVER be placed across the entire scrotum in an attempt to crush both spermatic cords in one go.

This technique requires more skill than banding and can cause serious pain and damage if not carried out properly.

Careful thought should be given as to whether it is necessary to castrate male lambs at all. Some people

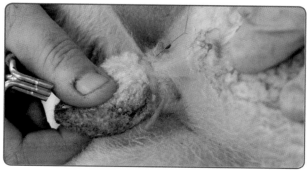

Above: A castration band.

consider it an unnecessary mutilation and prefer to carefully segregate male and female lambs before they mature to avoid unwanted breeding. Fast-growing breeds may reach slaughter weight before sexual maturity and therefore it could be argued that castration is pointless. At the time of writing (2015) castration and tail docking are still permitted procedures, subject to the rules mentioned earlier, but there is growing pressure in the European Union to outlaw both.

Tail docking is carried out mainly to minimise the risk of flystrike (see Chapter 6). A tail that is reduced in size is less likely to collect faeces – particularly when sheep gorge themselves on lush summer grass, leading to scouring – and is therefore less likely to attract flies. Hill breeds, however, are often left with their tails intact, and several breed societies also insist on this in pedigree stock. As with castration, you need to ask yourself whether you really need to dock. Welfare guidelines state: 'Tail docking may be carried out only if failure to do so would lead to subsequent welfare problems because of dirty tails and potential flystrike.'

If both tail docking and castration are deemed necessary, the advice is that both procedures should be carried out at the same time, to minimise stress and the risk of mis-mothering or abandonment.

Docking using rubber bands is similar, but much easier than castration. By law, you must leave enough tail above the band to allow the anus and, in ewe lambs, the vulva, to be covered.

Below: Tail being docked.

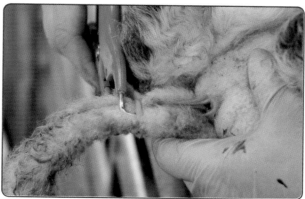

RAISING FOR MEAT

Although sheep are the ultimate multi-purpose animals, providing numerous useful products, the vast majority of those raised throughout the world are kept to provide meat. Lamb has always been a fairly expensive meat to buy in the UK. Although meat imported from other countries and sold at knock-down prices in the supermarkets may appear tempting to customers, there is nothing that can beat the taste of grass-fed, home-reared lamb, which has been allowed to grow at a natural rate, without the aid of supplementary feed.

Different breeds will, of course, grow at different rates. Primitive and hill breeds are naturally slow-growing, whereas the modern breeds such as Texels and Beltex – the mainstays of the sheep industry – have been selectively bred to produce meaty carcasses in a fraction of the time taken by their less-productive cousins.

This book is aimed largely at beginners to sheep-keeping who are likely to be raising lambs on a small scale for their own needs and maybe to sell meat to friends and neighbours, or at farmers' markets. Producing finished lambs on a more commercial level is a whole different ball game and beyond the scope of the *Sheep Manual*. Lambs sold at market to major buyers or direct to supermarkets have to be spot-on in terms of weight and size – largely because the meat has to fit the packaging. Yes, the size of the lamb chops you buy in the supermarket is dictated by the size of the polystyrene tray! If farmers present animals that are over the specified weight, they are financially penalised rather than being rewarded for producing a bigger carcass.

When you're producing meat for yourself, the nice thing is that you can choose how big your chops and various joints will be, as you can make your own mind up about what age

A QUESTION OF TASTE

As with all species reared for meat, there is some debate about which is best for flavour – males or females. Butchers will almost always advise you that ewe lambs and wethers (castrates) give better quality meat but, in practical tests, little difference has been identified, even when tasters have been given meat from older, uncastrated ram lambs.

More importance has been attached to how the sheep's diet can affect the end product. Research has been carried out, which shows that flavour can suffer if sheep are finished on legumes (including lucerne and white clover), brassicas, oats, maize silage and soya. Experts recommend that if such ingredients are included in the diet, sheep should spend at least the final week before slaughter with grass as their only source of food.

The reasoning behind this is that it affects the pH level in the muscle. A pH higher than 6.0 results in dark, dry muscle – often described as Dark Firm Dry meat (DFD) – giving a meat colour not favoured by the consumer.

to slaughter your sheep. It's entirely up to you. Whatever size you decide to grow them to, refer back to the section on condition scoring sheep to assess whether or not they've reached the minimum size for slaughter. There's no point in sending your animals to the abattoir if they're just skin and bone – what you want is some nice succulent meat with a decent coating of fat for flavour.

Lamb, hogget, wether, mutton...

Although we tend to talk mostly about 'lamb', sheep meat comes in various guises, with names changing according to age and, sometimes, geographical location. Generally speaking, the following terms apply:

- **Lamb** – from a sheep less than a year old. Flavour is mild, fat content is normally fairly low, and it doesn't require long cooking times.
- **Hogget** – slightly older lamb, killed between one to two years old, this has more depth of flavour, but may need hanging for longer after slaughter and longer, slower cooking. Top restaurants have long chosen hogget meat (or wether meat – from castrated males of the same age) because of the stronger taste and bigger size of cuts produced.
- **Mutton** – anything over two years old, and often specifically grown on until the age of four. Mutton has a much stronger flavour as well as having more fat. Like hogget, mutton benefits from hanging to allow the natural enzymes in the meat to help tenderise the meat and develop the flavour. Cooking needs to be long and slow, but the flavour is worth the effort. Joints are naturally larger and the meat is often chosen for curing.

'New season' lamb – meat on sale in early spring and often promoted heavily as the perfect family roast at Easter – has always commanded high prices, but what the average customer doesn't realise is that meat available at this time of the year will have been intensively reared on concentrates in order to achieve fast growth and almost certainly housed indoors.

Our ancestors chose to eat mutton rather than any younger sheep meat, largely because it was more practical and more profitable. Think about it – why kill a lamb under a year old when you can grow it on, breed from it, harvest its fleece, and still enjoy the meat later on? When you look at it that way, it makes perfect sense.

Intensive farming has changed the way we buy and eat. Chickens are often barely five weeks old when slaughtered; pork from modern cross-breeds can be from pigs as young as three or four months old. Only beef – the bovine equivalent of mutton – tends to be more mature when it reaches the shops and still wins in the popularity stakes over the younger, milder-flavoured veal.

Although mutton's mass following waned considerably after the end of the Second World War, it has seen something of a resurgence in demand in recent years, thanks to awareness-raising crusades led by some high-profile supporters. Prince Charles launched the Mutton Renaissance campaign back in 2004 and restaurateurs and celebrity chefs have been quick to support the drive to get mutton back on our dinner tables. Somewhere along the way, mutton gained an unfair reputation as a poor man's meat, often tough and chewy and vastly inferior to lamb. Now, however, it is becoming established once again as a gourmet meat to be savoured and enjoyed.

Planning for the abattoir

Finding an abattoir isn't as easy as it might have been a few decades ago. European regulations and requirements for costly improvements have forced the closure of countless small family-run slaughterhouses, so be prepared to search further than your own town to find your nearest one. The Food Standards Agency licenses all abattoirs in the UK, polices the way they operate, and employs vets on site to inspect animals before and after slaughter. A list of licensed premises can be found on their website, www.food.gov.uk, or ring your local office for more information. When you find the right place, book well in advance because there might be a waiting list at some smaller places. Few abattoirs will kill every day, and many dedicate specific days to certain species.

Check with the abattoir regarding any specific requirements, including drop-off time and when your meat will be ready for collection. Some abattoirs will insist that the sheep must be dry on arrival, so you may need to house them overnight. They will certainly expect their fleeces to be relatively clean and free from faeces. They may also want you to make sure that your sheep are as empty as possible, making evisceration – the process of removing the internal organs – easier and cleaner. This will mean restricting your sheep to just water and straw the night before slaughter.

Movement licences

In order to move your sheep, you need to use one of the systems mentioned in Chapter 2, filling out a movement licence, whether online or on traditional paper forms – whichever is the case in your particular area. As mentioned earlier, paper movements are still permissible in some circumstances, so check with the abattoir, your local authority, or the relevant government department as to the current situations. At the time of writing, for instance (2015), Wales is still waiting for the introduction of an electronic movement system; England and Scotland have theirs up and running, but are still accepting paper movements as farmers adjust to the systems. Don't forget to record the move in your own flock book, too.

Withdrawal periods and identification

Before you book your abattoir, make sure that any medication you have given your sheep will be well out of their bodies before they are slaughtered. Veterinary drugs have different withdrawal dates (i.e. the amount of time which has to pass before the animal can enter the food chain), so check the information on the bottle or pack, and note the details in your medicine records book. Only then should you arrange slaughter. You will need to fill in a Food Chain Information declaration confirming that – as well as satisfying various other health criteria – your sheep are clear of any medication.

Make sure your sheep are tagged in accordance with the current guidelines for your area. See Chapter 2 for a reminder.

CHECKLIST BEFORE SETTING OFF
- Are the sheep fit and well enough to travel?
- Are they correctly tagged?
- Have you completed all the relevant sections on the movement licence?
- Is your trailer roadworthy, clean, well ventilated, and the tailgate securely closed?
- Are the sidelights, brake lights and indicators working?

Arriving at the abattoir

Taking a trailer to a new place – particularly if you're not very experienced at towing – can be a bit nerve-racking, so try and arrange a trial run beforehand. Most abattoirs will be happy to let you practice going through the motions during a quiet time. If you're confident on the day, it makes the whole process less stressful for you and quicker and more efficient for the abattoir staff.

Some abattoirs have one-way systems where you can just drive in, offload and then drive out, but others will require you to reverse into the lairage – the offloading and holding area. If you don't have the time or opportunity to do a dummy run in advance, don't worry, because there will always be someone on site willing to do it for you – whether a member of staff or another customer waiting to offload.

On arrival, you will be met by a member of staff from the abattoir – possibly the slaughterman himself – and a vet acting for the Food Standards Agency. The vet has a responsibility to inspect the animal before and after it is slaughtered, to check the food chain information supplied, and to ensure that the meat is fit for human consumption.

Depending on the size of the abattoir and available facilities, you may be able to wash out your trailer on site, or you may be asked to sign a declaration saying you agree to wash out when you return to your holding.

Slaughter procedure

Once you offload your sheep they will be herded into a holding pen until the time of slaughter. Under UK law, sheep have to be rendered unconscious before killing, although exceptions are made in the case of religious beliefs for Jewish and Muslim communities. In conventional slaughter, either electrified tongs are applied to the temples or a captive bolt gun is used to fire a metal rod into the brain. Both cause the sheep to lose consciousness, and then the throat is cut and the major blood vessels are severed. This results in rapid blood loss, followed quickly by death.

If you want to get your sheepskins or horns back,

WHAT HAVE I DONE?

Taking your first animals to slaughter is never easy when you've spent months feeding and caring for them. It's important to remember why you got them in the first place and not to get too sentimental. Your aim was to produce your own meat in a healthy and ethical way, giving your sheep the best possible life while they were in your care. It wouldn't be normal if you didn't feel a slight pang of guilt when they trot off the trailer, but it does get easier the more you do it.

remember to ask the abattoir in advance (see Chapter 12 for more on using by-products). When the carcass is cut open for the stomach and internal organs to be removed, the vet on site will inspect them and decide whether they are fit for consumption. Sometimes abnormalities will be found – such as 'milk spots' on the liver (lesions due to worm larvae) – and you won't be allowed to have them back.

After being inspected, the carcass will be chilled and hung before being butchered. With lots of abattoirs, you could be dropping off your sheep on a Monday and collecting as soon as Thursday or Friday, whereas others will keep the carcasses for up to two weeks, which is much better for the quality of the meat and flavour. Older sheep benefit from a longer hanging time, but the abattoir may not have the space to hold on to them for you, so ask when you book your sheep in.

Briefing the butcher

Some abattoirs offer a full butchery service, while others will simply kill and hang the carcass, in which case you need to arrange delivery to a butcher. Whoever is cutting for you, make sure they have detailed instructions of how you want the job done. Each butcher will normally have a standard way that most carcasses are cut, so if you want anything special, do make yourself clear beforehand. You might, for instance, want your legs whole for a big family roast, or split into smaller joints suitable for two people. Similarly, you can choose between having chops or keeping the loin whole for roasting.

If you want the internal organs back, say beforehand or you might find they are automatically thrown away – or maybe the liver will end up in the butcher's window. Specify how you want everything packed, too. Half lambs are sometimes packed in cardboard boxes lined with greaseproof paper, with the joints laid out neatly but not bagged; some butchers bag each joint individually and pack the chops in twos or fours. If you don't mind paying extra, you may be able to have everything on polystyrene trays and over-wrapped with plastic film, or vacuum-packed – both of which are handy if you're planning to sell at farmers' markets (see page 132).

BUTCHERING A LAMB

There are lots of regional variations in the way butchers deal with a carcass, but most will start by cutting it into three 'primals' – the front (shoulder), middle (loin), and back end (legs) – for ease of handling. Some take the neck off first, while others prefer to leave it on until just before the front section is tackled.

1 The carcass is hung whole until ready to be butchered.

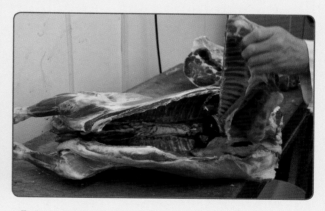

2 The neck is removed. This is often used for stews or casseroles.

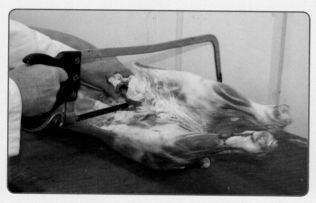

3 The legs are removed.

4 The breast is removed from each side.

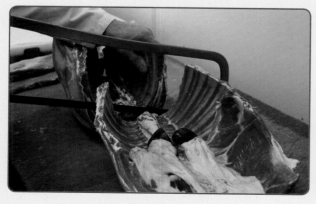

5 The shoulder section is separated from the loin.

6 Incisions are made either side of the spine in the shoulder section and the ribs are removed.

7 The shoulders are separated.

8 The loin section is split into two.

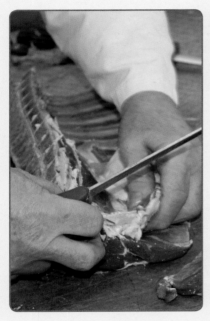

9 The loin is trimmed of excess fat.

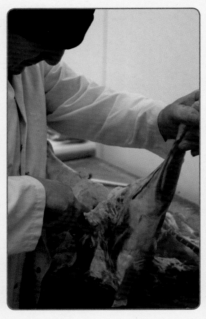

10 The legs are separated.

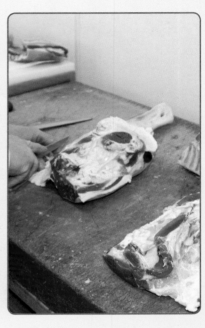

11 Excess fat is removed from the legs.

12 Chump chops are sliced from the top of each leg. The chump – often described as the equivalent of fillet steak – can also be cut as a small joint suitable for roasting.

13 The loin is cut into chops – or it can be left whole as a roasting joint.

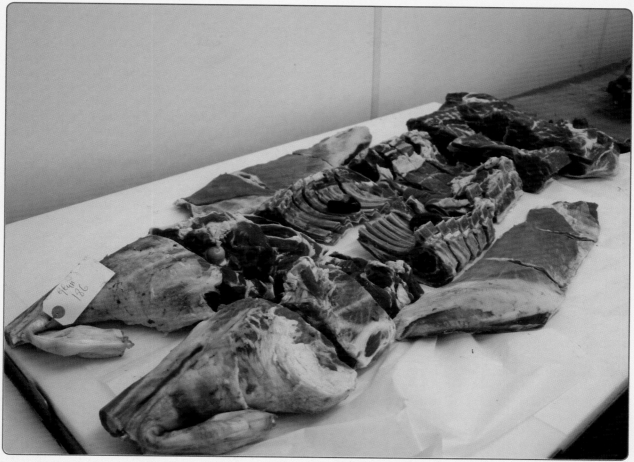

14 The finished job – a whole lamb ready for packing.

Home slaughter and butchery

For most people, the slaughter of livestock is best left to the professionals, but there are some reasons why you may want to do your own killing. Some supporters of home slaughter say that if they are prepared to eat an animal they have raised, they should have the courage to kill it themselves, too. There is also the argument that killing on site instead of transporting animals to an abattoir that may be several hours away reduces stress. Part of the reason for wanting to do this involves the welfare of the animal, but it's also been shown that stress can have an adverse effect on the quality of the meat.

Home slaughter is perfectly legal if carried out correctly, but there are a number of things you must bear in mind, should you choose to go down that route. Firstly, and most importantly, you need to be confident that you can carry out the slaughter quickly and humanely and in a way that reduces suffering to an absolute minimum. The Humane Slaughter Association is your first port of call. A registered charity, it not only lobbies for higher welfare standards at abattoirs but also provides some excellent training courses and educational materials aimed at helping livestock producers, large and small, to understand the procedures involved. Resources include online guides, printed publications and DVDs. Details can be found on their website, www.hsa.org.uk

From a food safety point of view, home slaughter seriously restricts what you can do with your meat. The law says that only you and the people who are regular members of your household can eat the meat – it can't be sold or given to anyone else, nor bartered for goods or services. So, for instance, if you had a guest cottage or a campsite and had holidaymakers visiting, you could not allow your guests to buy any of your meat.

Waste disposal is also an issue with home slaughter, as all by-products – blood, skins, bones, internal organs and so on – must be disposed of properly. You can't simply stuff everything into rubbish bags and expect the binmen to take them away. Everything must be properly incinerated – for instance by a fallen stock company – or collected by a business registered to use animal by-products as animal feed, which may include local hunts, maggot farms and sometimes zoos.

Selling your meat

The Food Standards Agency website (www.food.gov.uk) provides plenty of useful information on the current legislation regarding food hygiene and selling regulations. Anyone who plans to sell their own produce must first register as a food business with the local authority's environmental health department. Officials will want to know information such as what you feed your sheep, how feed is stored and where you store your meat between collecting from the abattoir/butcher and selling to customers. They will want to inspect your storage facilities, examine fridge and freezer records, and see copies of any food hygiene qualifications you hold as well as your HACCP (Hazards and Critical Control Points) plan, which outlines how you seek to avoid health risks in the course of your business.

Having your sheep professionally slaughtered, butchered and packed at an abattoir – or killed at an abattoir and then cut and packed at other licensed cutting premises – is by far the simplest way of starting off your meat venture. Doing it this way allows you to sell the meat to others. However, if you choose to carry out butchery yourself with a view to selling to the public, you will have to set up your own butchery premises, which must meet stringent hygiene standards – much higher standards than those for merely storing and selling meat packed by a licensed third party. It can be a costly business, kitting out a room or building to the appropriate standards, and you will be under far more scrutiny than a mere retailer who doesn't handle raw meat.

Selling at farmers' markets

Originally an American idea, farmers' markets have really taken off in the UK, growing in popularity and number all the time. Celebrity chefs and the proliferation of 'foodie' TV programmes have done much to encourage the public to sample more local produce and choose quality over cheap factory-farmed products. Recent years have also seen several food health scandals, such as the one involving horsemeat being mislabelled as other meats, and traceability of food is becoming a much bigger issue for shoppers than ever before.

There are hundreds of farmers' markets – often labelled 'rural markets' or 'community markets' – in the UK, with a large proportion belonging to the National Farmers' Retail & Markets Association (FARMA), an organisation set up to represent producers selling directly to the public. FARMA certifies farmers' markets in the UK, which operate under its guidelines. Certification means they have been independently inspected and meet certain standards – stallholders must be local farmers, growers and food businesses selling their own produce.

You may find farmers' markets organised by farmers' co-operatives, local authorities, community groups, or private individuals. Getting into an existing one can be tricky unless you have a niche product, as organisers don't like having more than one stall selling the same thing. Although rules will vary, most markets have the same kind of ideals:

- They exist to enable local farmers and producers to sell direct to the public; to give consumers the chance to buy fresh, locally grown fruit and vegetables, locally reared meat and home-made products and to help preserve the rural economy.
- Producers must have grown or produced. The term 'producer' includes the stallholder's family and employees when they are directly involved in the business. Stallholders must not sell products or produce on behalf of, or bought from, any other farm or supplier – so ensuring complete traceability.
- Produce must be from a defined 'local' area. 'Local' is usually taken to mean within 30 to 50 miles of the market. However, producers from further afield may be considered if the produce they are selling cannot be sourced within the specified 'local' radius. In the case of applications for pitches by producers of similar foods, preference is normally given to the most local producer.
- Details of where meat was raised/processed should be on all labelling and producers should display trading names clearly on their stalls, together with a contact address.

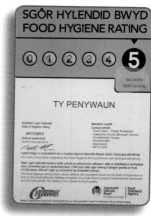

- Stallholders must comply with all local and national laws and regulations regarding the production, labelling, display, storage and sale of goods. All producers must observe the current food hygiene regulations (see the FSA website). The producer's food hygiene rating (awarded by the local authority following an inspection of premises) must be displayed.
- Producers will need public and product liability insurance certificates. Public liability insurance protects against claims by a third party injured or damaged as a result of your business, e.g. if your stall falls down and hurts someone. Product liability insurance protects against claims arising from the actual food you are providing.
- By law, prices must be clearly displayed – either on the pack or prominently on the stall. This makes sense from a practical point of view, too. Having to ask for prices is off-putting to customers, and will scare some away.
- Most loose foods must be sold by net weight, using approved metric weighing equipment. If the food is pre-packed, the metric weight must be marked on the pack, but you can also add an imperial weight in a less prominent position. You must have accurate weighing scales, which are calibrated for metric weights and approved by your local Trading Standards officer. Spot

checks of your scales can be made at any time either at the market or at your farm.

■ Producers 'adding value' to primary local produce (e.g. by processing) should use local ingredients.

Why sell at markets?

Selling direct to the public in this way can be very satisfying if you enjoy talking about your produce, your farm and your lifestyle. Getting feedback from happy customers is always good and may give you ideas for future products and ventures. Markets can be good 'loss-leaders', too. For instance, if your customers like your sausages and burgers, they may want to buy a half or full lamb from you at a later date.

Lots of producers selling premium products prefer to trade at markets in order to cut out the middleman, so increasing the profit margin. Another advantage over simply waiting for customers to turn up at your farm gate is that most markets are only open for a few hours, so your selling activity is concentrated.

Cost-wise, it's not an expensive thing to do, unless you have your heart set on a premium site in an affluent area. You'll have to pay for the cost of your table or gazebo space and, if you need power to run a fridge, there may be an additional charge for electricity. All in all, it works out a lot cheaper than kitting out your own farm shop.

The downside is that markets can be a big commitment. Being a constant feature is vital, and the organisers won't be happy if you don't turn up on various occasions. Regular customers, too, will be disappointed, and may decide to look elsewhere.

A huge amount of time and planning goes into getting goods ready for sale. In fact, if you costed in your time, you'd probably decide it wasn't worth doing it! If you're selling fresh meat, rather than frozen, you'll have to ensure that your animals will be ready for slaughter when the markets need them. This means working even further backwards to when your animals need to be born or bought in. You'll also need to have a contingency plan in case you run short of stock – for instance, if you lose any animals through sickness, or if they don't reach the required weight in time.

Factor in time for processing, packing and labelling and also think what you'll do if you don't sell everything. Do you have an outlet for surplus stock? Frozen meat doesn't sell as well as fresh, so look at alternative ways of offloading your spare produce.

Tricks of the trade

Making your stall stand out from the rest is an excellent way of attracting potential customers.

■ Tell the story behind your produce. Include photographs and information about your livestock and the food preparation process. If you have won any rosettes for your produce or livestock, show them off.

■ Allow plenty of time for setting up, as first impressions are important.

■ Label everything clearly – to comply with current legislation, and to make browsing easier for your customers.

■ Think about your appearance, smile and be welcoming. Coats or aprons should be clean and smart to give a professional look. Hair should be tied back or tucked away under a hat, and don't forget to make sure your hands and nails are clean.

■ Work hard to engage potential customers in conversation. Tell them about other farmers' markets you attend, and give out flyers and business cards.

■ When talking to one customer, keep an eye out for others who might be waiting to make a purchase but feel they can't butt in because you're talking. Don't lose a sale by talking to one customer for too long.

■ Don't over-price your meat, but don't undervalue it, either. Do plenty of research beforehand and find out what others supplying similar products are charging.

■ Draw attention to yourself in the media. Think of ways to get news organisations interested in your business. Are you the first producer to be selling a certain type of food at your farmers' market? Are you raising unusual or rare breeds? Has your business won a grant, which has allowed you to boost production? Have you taken on new staff? Have you won an award? Do you have a celebrity customer who loves your meat?

■ Don't be dismayed if you don't sell out every time. Everyone gets good days and bad ones – and, chances are, if you've not sold much at a particular market, your neighbouring stallholders won't have, either. Shoppers can be swayed by the weather and distracted by other local events staged at the same time as the market. Don't give up immediately – hang on to the same stall in the same place, and keep providing quality food and you'll soon start to see the same faces coming back time and time again. And that's one of the things that make all the hard work and commitment worthwhile.

Louise Fairburn, The Rigsby Flock of Lincoln Longwools.

MAKING THE MOST OF YOUR SHEEP

Louise Fairburn loved her Longwools so much, she had her wedding dress made from their wool and they even took part in her special day.

Why do so many people keep sheep? Two reasons normally spring to mind – meat and wool. But there is so much more to be derived from this amazing animal – as generations of shepherds would testify – that it has to rank as the most valuable all-rounder in the world of livestock. You can eat it, wear its wool and skin, use its milk, and turn other by-products into a whole host of everyday items that we tend to take for granted, as well as even more luxury ones.

Think, for instance about lanolin – the sheep's natural waterproofing substance. It's used for goods as diverse as cream for sore nipples and lubrication for engine parts. It's used in make-up and beauty products as well as household paints and it's even used on oil rigs to ward off corrosion. The estimated one billion sheep in the world may be owned mainly by large-scale farmers with flocks of immense proportions, but there are a growing number of smallholders, hobbyists and artisan producers who are quietly making a living – or, at least, ensuring their sheep pay their way – by using precious by-products in creative ways.

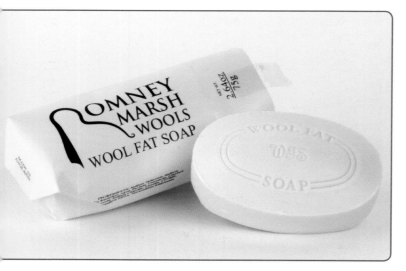

Curing lamb and mutton

The gastronomic qualities of lamb and mutton are well known, but have you ever thought about preserving these meats by curing, in the same way as you would pork? Our ancestors would traditionally have killed livestock as the autumn approached and then built up a store of all kinds of meats – including lamb and mutton – for use over the colder months by salting and hanging it from the rafters, often above the hearth.

In various countries across the world, religious beliefs rule out pork, making lamb, goat and chicken the meats of choice. Not surprisingly, lamb appears in many different guises on international menus, from traditional-style sausages to salami, and from ham to pâté. More and more adventurous home curers are reaping the benefits of making their own air-dried charcuterie, jerky and biltong-style snacks. None of this is difficult to do, but you do need the correct storage conditions and ways of controlling temperature and humidity.

Curing follows pretty much the same procedure, whatever meat you use. There are basic methods, wet and dry, but the aim with each is the same – to get salt absorbed into the meat in order to lengthen its shelf life. Dry curing is done by rubbing in the salt; wet curing by immersing in a salt solution (brine).

Above: Lamb jerky.

While you're practising, use less expensive cuts first, such as the breast, so that if it goes wrong, you haven't wasted too much. There's a saying about curing – 'It's not a sin to bin'. So don't worry if it doesn't work out perfectly the first time.

Probably the easiest way to start is to buy a ready-made curing salt mixture from one of the many specialist suppliers, and then add your own additional herbs and seasonings to suit your own tastes. Most curing mixes will include small quantities of sodium nitrate to kill bacteria, help speed up the curing process, and preserve the pink colour of the meat. In traditional cures, potassium nitrate (saltpetre) was added to salt and, although harder to find these days, it can be bought on the Internet.

Below: Lamb sausages.

RECIPE FOR LAMB HAM

This is a very simple method for dry-curing a lamb ham. Bone in or bone out is up to you. If you leave the bone in, make sure to work the cure right into the small gaps around the bone on the surface of the joint.

Choose a whole leg of lamb or mutton or a leg joint – size doesn't matter, but make sure that, if you're planning on curing in a fridge, you have space for storage.

Weigh the leg and measure out the correct amount of ready-made or home-made cure. With ready-made cures, the instructions will be on the packet, but expect to use about 30g per kg of meat. Have extra cure ready, just in case you use up more on one particular bit; don't leave any part of the meat uncovered.

The flavourings you add into your cure are entirely up to you, but you might want to include some brown sugar, some rosemary, bay leaves, crushed peppercorns, crushed juniper berries, cloves, mace, garlic, or ginger. For a 1kg leg or joint, around a teaspoon of each should do, but feel free to add more or less, depending on your preferences.

- ■ Combine all the ingredients in a bowl, add the salt cure and mix well.
- ■ Place the joint in a non-corrosive tray and rub the mixture in well, making sure to get into any small folds or openings.
- ■ Wrap the joint tightly in cling film or seal in a bag using a vacuum packer. Put it in a fridge or very cool place – no warmer than 6°C. To work out the curing time, allow one day per 13mm depth of meat (measuring from the thickest point) plus an additional two days.
- ■ Turn the joint every other day.
- ■ When the curing time is up, unwrap and rinse the excess cure and seasonings from the meat. You could, at this stage, dry off the ham, wrap it in muslin, and find a suitable place to air-dry it for several months, depending on storage facilities available.
- ■ Cook as you would a pork ham, boiling or roasting – or both. Some people prefer to parboil and then finish in the oven.

Dairy produce

Still way behind goats' milk in the popularity stakes in the UK, sheep milk is big business in many parts of the world – and with good reason. As well as being more productive all-rounders than goats, and easier to handle and maintain than dairy cattle, sheep produce milk that is much higher in calcium than either of the other species (162–250mg/100g, according to breed, as opposed to 103–203mg for goats and 110mg for cattle). Protein levels are higher, too, with sheep at 5.6% of total solids, compared to 2.9 for goats and 3.4 for cattle, and it's also richer in vitamins A, B and E.

The number of flocks producing ewe milk for sale to others or for use in artisan products has grown significantly in the UK in the past three decades, partly because of people with food intolerances seeking an alternative to cow and goat milk. In 1983, the British Sheep Dairying Association was formed to promote the production, manufacture, marketing and consumption of dairy products, and to help those setting up new dairy ventures.

BEST DAIRY BREEDS

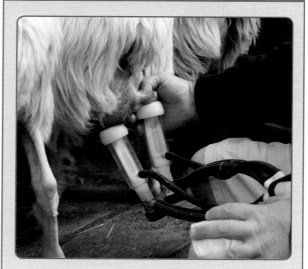

Any ewe in lactation can be milked, but, as with any livestock, there are certain breeds that are more suited to particular jobs. Popular breeds used for milk production include: East Friesland, Lacaune, British Milksheep, Dorsets and Zwartbles.

As with dairy cattle, the preferred breeds tend not to be as hardy as meat breeds. It's important, therefore, to provide access to housing or shelter all year round, and some flocks are allowed only limited access to pasture during the day and returned to their sheds overnight.

Sheep will be milked twice a day and will normally continue in lactation for around 245 days, compared with approximately 280 for goats and 305 for cows. Actual yield of milk is miniscule compared with goats and cows. According to the British Dairy Sheep Association research, during one lactation, sheep can be expected to produce as little as 350kg, whereas goats might manage more than 1,000 and cows around 8,500.

Keeping a ewe milking means breeding from her, but what happens to her lambs if you want the milk for yourself? Various systems are operated:

- Lambs stay with the ewe for 24hrs to four days to make sure they receive sufficient colostrum, after which they are bottle-reared;
- The lambs stay with the ewe for a month, during which time no milking is carried out; or
- After the first 24 hours, the lambs are given 'controlled access' to the ewe, so that she can be milked for profit, but the lambs can still continue to suckle occasionally; the rest of the time, they are fed artificially. They could, for example, be bottle-fed throughout the day, but allowed back to the ewe at night.

Above: Sardinian Pecorino cheese.

Above: Slovak Bryndza cheese.

Above: Feta cheese from Greece.

Cheese made from ewe milk is made in several parts of the world, and there are more than 80 recognised types. Pecorino (from the Italian 'pecora', meaning sheep), feta, manchego, Roquefort and ricotta are among the best known; producing ricotta is a resourceful way of making use of a by-product of the conventional cheese-making process – whey – which would otherwise be disposed of as waste. The name 'ricotta' actually means 'cooked twice', as the whey is put through a further production process to make it ready for consumption.

Ice cream, yoghurt and smoothies made from ewe milk are popular, too. There are a growing number of companies devoting their resources to making delicious desserts using this nutritious alternative. The richness of the milk makes it perfect for ice-cream making because, unlike ice cream made from cows' milk, no additional cream, butter, or egg needs to be added.

A few specialist ice-cream parlours exist, and mobile units are becoming a more familiar site at outdoor events like agricultural shows and music festivals.

Below: A sheep ice cream stall at an outdoor event.

Shepherds Ice Cream

Ken McGuire

Non-edible products

There's lots of scope for making money out of the meat and milk of the sheep, but its woolly coat offers plenty of opportunities, too.

Wool, of course, has historically played an important part

Above: Black Welsh Mountain sheep breeder, Oogie McGuire, working with her own wool.

in industries across the world, and, after a few decades of losing out to man-made fibres, the demand for this durable and versatile, eco-friendly product is growing once again – not only for clothing, but for carpets, upholstery, building insulation and mulch for flowerbeds.

In recent years, the art of felting (below left) has become a popular pastime – and a business venture for some. Felting involves bonding and shrinking wool fibres together to create a dense cloth. There are various types of felting, but wet felting is probably the easiest and most popular, using warm soapy water, agitation and compression to encourage wool fibres to stick together to form a piece of fabric.

When you take your sheep for slaughter, you can ask for your skins back so that you can have them tanned. Professionally tanned, attractive fleeces can be worth quite a bit of money – sometimes more than the meat will sell for.

Comparatively few people do ask for their skins back and, for a long time, the EU made it quite difficult for small-scale

www.lizmangles.co.uk

breeders to do so. The abattoir will normally sell skins on to a commercial processor, so they may charge you a small amount if you are going to have them back. As skins are classed as animal by-products, there is some inevitable red tape involved, so ask your abattoir or contact Defra, or the appropriate devolved administration, for the up-to-date paperwork you need to complete regarding handling, transportation and storage.

If a sheep has a shoddy fleece – maybe damaged by flystrike or injury – when it goes for slaughter, don't expect it to produce the perfect skin once it's dead. Younger sheep will produce the best skins. Timing is also important because of the way the fleece moults and starts to regrow; if you want skins from lowland breeds, they will need to be slaughtered before the middle of October at the latest; with hill breeds, the beginning of November is recommended.

Skins need to be salted within four hours of slaughter, and the abattoir will normally do an initial salting if you ask. There are a few specialist tanneries that will do the job for you, and, unless you are delivering straight to them, they will send you full instructions regarding how the skins need to be salted, dried and packed before sending to them.

Creating a niche or novelty product can be a great way of maximising profit from your sheep. Sculptor Liz Mangles, who runs Ewenique Furniture in Portland, Dorset, has developed a range of novelty sheep foot stools (above), which are made of hand-carved lime wood, then upholstered and covered with fluffy sheepskin. Some can be ordered in specific rare-breed fleeces, or you can even have one wearing a fleece from your own flock.

Below: Tanned sheepskins.

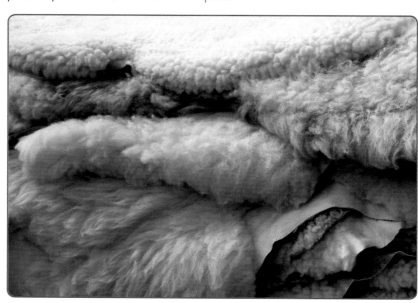

SHOWING

At some point, when you've found the breed you love and have put time and money into breeding a top-quality flock, you may get the showing bug.

To those who have never tried it, it may seem like a load of nonsense, with serious-faced people in starched white coats and shirts and ties primping and preening their animals like they were prize pooches at Crufts. 'What's the point?' is often the question asked. Well, the point is that everyone, whether breeding pedigree or not, should be breeding from the best to produce the best. We're not talking simply cosmetics here – although every breed of sheep has its own particular characteristics, which have to be just right, and each animal registered as pedigree must have all the physical attributes that make it a good breeding animal. In short, each beast has to be functional and suited for the purpose – as well as being good looking in other ways.

Agricultural shows are a way of showcasing the best and there is no better place to learn about what makes a good example of the breed. Taking part can be a thoroughly enjoyable activity and there are many thousands of people who do it, year in year out, building up a network of friends from all over the UK.

Making a start

It may be that another breeder – possibly one you bought your original stock from – persuades you to give showing a go. To make single-breed classes viable, there have to be sufficient exhibitors, otherwise breeds will be amalgamated into mixed classes, which are sometimes deemed as unfair to particular breeds and can make the judge's job more difficult. Bigger classes also mean that, when you win, the thrill of being given that rosette is better than ever.

If you haven't been cajoled into competing by someone else, but fancy showing for the first time, it can take a bit of courage to fill in that first entry form, but don't let fear of the unknown get the better of you.

If you talk to regular competitors, there are numerous reasons to enter. The prize money won't make you a fortune, but the satisfaction makes up for it. And, of course, you are putting your stock centre stage. There's no better way of advertising your flock to potential buyers than parading them at a high-profile show and coming out of the ring with a top judge's seal of approval.

Who can show what – and where?

Showing livestock is different to showing some pedigree pets, in that you don't necessarily need to work your way up through small regional heats in order to compete at major shows. As long as your animals meet the specified criteria, you can choose wherever you want to exhibit. Lots of people start at small, local shows and gradually get more ambitious, but there's no reason why you can't go straight in at county level. Lots of people do. Smaller shows are more laid-back

and will normally allow non-pedigree exhibits, while the prestigious county shows are showcases for the breed societies, are much more competitive, and animals will need to be registered in order to take part.

You don't need to have bred the animals yourself, although some of the more experienced breeders prefer not to show anything they've bought in. The fact that they are registered in your name is all that matters.

Although a lot of people see farming as a rather male-dominated profession, livestock shows are a good way of proving just how many women are involved in agriculture. Children are encouraged, too, and every show will have classes for young handlers, although most have minimum age restrictions.

Finding a show

Breed societies compile lists of forthcoming shows and individual shows will have news on their own websites. Make sure you start your research in plenty of time, because livestock schedules – the booklets that list the classes you can enter, fees payable, details of judges, prize money and so on – are sent out months in advance and entries have to be in early so that catalogues listing the classes and details of exhibits and owners can be printed.

You will need to fill in forms with details of your animals and their pedigrees, as well as other information, such as how many pens you require. You will probably need to book a camping space or other accommodation; some shows

offer simple rooms in very basic accommodation blocks, but most exhibitors prefer to set up camp with friends, often staying in their livestock trailers – after a thorough washing out, of course!

Below: Preparing for all weathers: an exhibitor fits a tarpaulin over his makeshift hotel room.

Shows and movement regulations

As mentioned in Chapter 2, moving any livestock on to your holding triggers a standstill on movement of stock already there – six days in the case of sheep. If you're planning to buy new stock, or hire in a ram, make sure the movement doesn't stop you going to your show.

Also, when your show team return, don't forget the rest of your livestock will be subject to restrictions (six days for sheep, goats, and cattle; 20 days for pigs). Anything going for slaughter can move as normal, but farm-to-market or farm-to-farm movements have to wait. This is where having an approved isolation unit can be useful (see Chapter 2).

What can you show?

Everything depends on what the show schedule says. Normally, larger shows will have numerous classes, for instance:

- 'Aged' or senior ram – often described as 'two shear' or over, which means it has been shorn twice
- Yearling/shearling ram (shorn once)
- Ram lamb
- Ewe (some say 'two shear', some may specify she has reared lambs in the year of the show)
- Yearling/shearling ewe (shorn once)
- Ewe lamb
- Group classes. These can vary in what is required – maybe three lambs sired by the same ram, or possibly a ewe with two of her offspring. Some shows have classes for a ewe and a single lamb. Read the rules carefully, as there will be differences between shows.

Some breeders take showing extremely seriously and 'do the season' – entering practically every show they can, even if the shows are hundreds of miles away. It can be a huge commitment, as well as being expensive and time consuming, and it's incredibly hard work, too. Keeping a sheep in 'show condition' for months on end involves real skill. When you're just starting off, it may be best to choose one or two shows that are fairly close together and aim at getting your sheep into peak condition for those.

Forward planning is the key. Experienced breeders carefully plan their births to suit the various age classes, so it pays to find out what those classes are and to think ahead. Animals are normally judged in separate breed classes first. The winners of the various categories compete for the male and female championship titles, and then battle it out to be breed champion. The breed champion will then go forward to the interbreed competition.

DON'T FORGET ...

- Make sure that vaccinations are up to date. Shows can be a minefield when it comes to disease; so don't take a chance by taking unprotected animals.
- Check that tags and tattoos are in place and legible.
- Set up your movement licence online – and remember to confirm the movement at the other end.

Choosing stock to show

Information on the breed standard can normally be found on the breed society website or in members' handbooks or newsletters. As mentioned earlier, only sheep that have satisfied the requirements of the breed standard should be registered and, when it comes to show time, each animal competing will be judged against the points specified in this standard.

If you think you might have a few sheep that are up to show standard, spend time watching them, because some have more natural 'presence' than others and will catch your eye. If you can, ask an experienced breeder to give you an opinion.

Preparation work

Depending on the breed you are showing, you may be required by your breed society to shear early in the year – often after 1 January. Many breeds are shown 'in the wool' – meaning they haven't been shorn. Check early on which applies to your breed, otherwise your show team's fleeces won't look as they should at the time of the show, and you won't stand a chance of winning. Part of the reason behind the early-shearing rule for some breeds goes back to the fact that numerous large commercial flocks are housed intensively over winter or at lambing time, and shearing is carried out to reduce heat stress when sheep are kept in large numbers in confined spaces.

Bear in mind that, if you are required to shear early – at a time when the weather may be at its worst – your sheep will have to be housed for several weeks until their fleeces begin to grow back.

Have a good chat with your breed society representatives or take advice from someone who is an old hand at showing your breed and make sure you get it right.

Unless your sheep are exceptionally clean, you may need to wash them two or three weeks before showing, to allow time for the fleece to recover. Lanolin – the fleece's natural conditioner and waterproofer – will be removed during

washing, but there are specialist 'show dip' products available to restore condition.

Washing can be a stressful experience – particularly for first-time show animals – so be prepared for them to panic. You will almost certainly need to put them in a head restraint of some sort. Professional livestock shampoos are available, but a mild baby shampoo is a good alternative. Some breeders go for the low-cost option – washing-up liquid or the cheapest shampoo they can find – but it's always a possibility that a particular sheep will have sensitive skin, which reacts badly, so do a skin test before going too far. Always ensure that whatever you're using is kept away from the eyes.

Some breeds will also apply a colouring product – often known as 'bloom' – after washing. This gives a yellow to brown hue and makes the sheep stand out (see below, left). Again, check with the breed society to see what is expected and acceptable.

Trimming sheep for showing is a real art form, which takes time to perfect. The experts have their own sturdy trimming stands, which they take to shows to complete the final

touches. Alternatively, a head restrainer is a cheaper and more portable option and can be fixed to a sheep hurdle.

When you're satisfied your sheep is safely restrained, you can start the final preparation process, which involves grooming with special metal 'carding' combs for 'carding up' the fleece. Normal practice is to start at the neck, working towards the back end of the sheep. Any soiled or straggly bits of wool need to be trimmed off at this stage. The aim is to enhance (or give the illusion of) a straight-backed animal with a solid, squarish frame.

Exhibitors often repeat this preening process several times before going into the show ring. Rams may also have their horns sand-papered and oiled to appear smooth and shiny.

Of course, after putting so much hard work into getting your exhibits looking good, you want them to stay that way, so fitting them with a coat – either purpose-made or made out of spare fabric – protects the fleece when they lie down in their pens. A final fluffing-up and maybe a fine spray of water, hair setting lotion, or oil just before the competition, and your sheep are ready to go.

Show ring practice

It's important to do some practice with your sheep at home in order to get them used to wearing a halter and walking obediently. This is probably the most unnatural thing a sheep could be asked to do, as they are naturally herd animals, not used to being on their own, and also because their natural instinct when approached by a human (unless food is clearly on offer) is to run away. The last thing you want when the judge approaches is for your sheep to turn on its heels and bolt.

There are different means of halter training. The traditional way, for generations, has been to fit the halter and tie the sheep to a gate or post until it stops struggling. However, this is now considered too harsh, as it can cause a great deal of stress and result in injury.

Much more preferable is the kinder, more gradual approach, similar to popular methods used in training dogs and horses. The first step is to get the sheep used to feeling a soft rope halter on its face but without securing the end of the rope to anything. Once the halter has been accepted, attach a lead rope and gradually start exerting pressure, pulling the rope towards you. If the sheep responds by moving towards you, release the pressure – so you are rewarding it for doing what you want. If the sheep pulls away, keep hold, but don't increase the pressure.

Once both you and your sheep are comfortable with the way the halter works, you can set up a makeshift ring using hurdles or straw bales and practise your ringcraft. The more you can watch other exhibitors in competitions, the better. You can pick up a lot of tips just through observation.

Getting ready for the show

Check that your trailer is safe and roadworthy. Bear in mind the transportation regulations mentioned in Chapter 2 and make sure you take any necessary documentation with you.

Whatever the duration of the show, you will need a basic survival kit to make life easier:

■ A kit box – a sturdy, wooden box or a plastic toolbox on wheels, in which to keep all your essential bits and pieces. If sufficiently well made, it will also double-up as a spare seat in the kit pen.
■ Bags of feed plus food and drink containers. Depending on the show, you may also need to take your own straw for bedding, so check with the organisers beforehand.
■ Grooming kit (shampoo, cloths, brushes, carding combs, hand clippers, etc.) for the final wash and brush-up.
■ Shovel and brush, for cleaning out the pens each day.
■ Mini tool kit – including staple gun, hammer, string, cable ties, etc., for putting up an informative display about your herd – and any rosettes, of course.

■ Business cards and flyers to advertise your stock.
■ Notebook and pens – for contact details of potential buyers.
■ Your exhibitor's coat and any handling equipment.
■ Small bulldog clips – to attach your exhibitor number to the pocket of your coat.
■ Food and drink – for you, not the sheep. There can be a lot of hanging around, so it's best to be prepared. You might not be able to leave the livestock building to get something to eat, so make sure you have something at hand.

Arriving at the showground

You will be expected to hand over a printout of your movement documentation on entry to the showground, or when you reach the sheep section.

You should have been sent a plan of the showground when you received confirmation of your entries, but if not, there will be stewards to direct you. Be aware that the later you arrive, the busier it will be, so if you're not too confident reversing a trailer, plan to get there in plenty of time. After off-loading, you will be sent to wash out your trailer. There will be a designated area for doing this, and, at bigger shows, there may even be staff to do it for you.

Keep an eye on proceedings

Get yourself a show catalogue, so you know what time judging is due to start, but be prepared for delays. Some judges take longer than others in their deliberations and schedules can slip. It may be frustrating, but don't be tempted to wander off, otherwise you might miss your class. Sometimes show PA systems aren't the best and can only be heard in certain areas, so keep an eye on other exhibitors who you know are in the same classes as you and be ready to move when you see them moving. If the show has good stewards, they will come and get you if you haven't already made your way to the ring. While you're waiting, use the spare time to give your animals a final spot of grooming.

Show ring etiquette

Make the effort to look the part and dress appropriately for the occasion. A clean white coat, shirt and tie, and smart trousers are the accepted dress and show you have respect for the proceedings and the people involved.

When in the ring, always keep an eye on the judge and be ready to present your animal as best you can when requested. All exhibitors will be asked to walk their animals around the show ring so that the judge can see how they move, and this is where good halter training really pays off. Afterwards, exhibitors will be asked to line up to have their animals inspected. The judge will study various aspects of the sheep, including conformation, the teeth and the correctness of the mouth, and he or she will assess whether it meets all the criteria set down in the breed standard. You will probably be asked a few questions, and the more knowledgeable you are, the better. While you're waiting to be inspected, don't be tempted to chat to neighbouring exhibitors or engage in conversation with spectators, as it's regarded as bad form.

If you get a rosette, remember to shake the judge's hand and thank him or her. If you don't win, be a good sport and applaud those who do. In some breed classes, competition is incredibly tough, so don't be too disappointed, and remember that everyone has to start somewhere. Don't be afraid to ask for feedback if you don't win anything, as it's always useful to know how to improve next time round. A good time to do this is after the competition is over, as the judge will normally visit exhibitors in the sheep lines or join them for a drink and a chat.

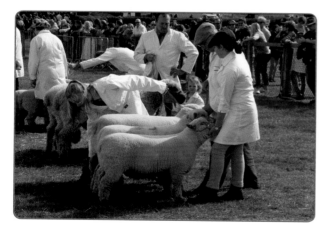

After the competition

For many showmen and women, taking part in an event is just as enjoyable as winning. Livestock exhibitors are an extremely sociable bunch and you'll soon build up a network of friends who share your passion for sheep. The larger shows attract participants from all over the UK and you'll soon get to recognise the faces of the regulars. Socialising with other competitors is a great way of learning more about your breed and making contacts that are useful when sourcing good stock – and, of course, it will help you keep abreast of all the important news and gossip!

The most welcoming shows will offer a 'stockmen's supper' to which all exhibitors are invited. There may be a nominal fee, or it may be a complimentary dinner thanks to a sponsor. However, even if there isn't anything organised, there will always be a bar on site packed with fellow exhibitors, as well as a few parties in the livestock sheds and the trailer park.

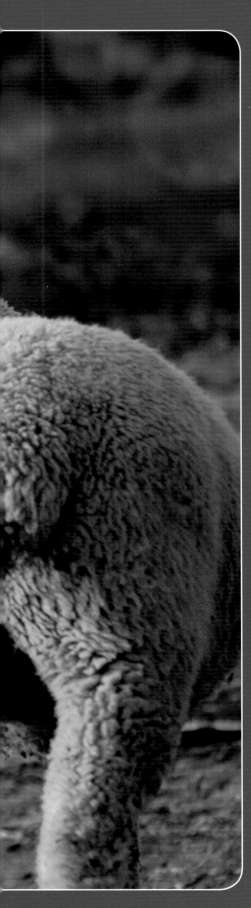

APPENDICES

Appendix 1

GUIDE TO SHEEP BREEDS

Listing each sheep breed in the world would make an entire book of encyclopaedic proportions, but this extensive catalogue, which includes the most popular breeds – and some of the rarest – should stand you in good stead for doing further research and finding the breed that suits you best.

Some breeds have the letter 'R' in red next to their name, which denotes they are regarded by the conservation charity, the Rare Breeds Survival Trust (RBST). The RBST publishes an annual Watchlist, which is an inventory of rare breeds in the UK – not just sheep, but cattle, goats, pigs, horses and ponies. They are categorised

according to the number of registered breeding females. With sheep, the categories are:

1) Critical – fewer than 300
2) Endangered – 300 to 500
3) Vulnerable – 500 to 900
4) At risk – 500 to 1,500
5) Minority – 1,500 to 3,000

The information here is based on the 2015 Watchlist. For more information, see the RBST website, www.rbst.org.uk

Debbie Kingsley

Badger Face Welsh Mountain

There are two types of Badger Face, the Torddu (which, in Welsh, means 'black belly') and the Torwen ('white belly'). The Torddu is predominantly white, with a dark underbelly and dark eye stripes, while the Torwen has almost reverse markings. Relatively small, with ewes between 40 and 50kg, it is a prolific breed that lambs easily and does well in harsh upland areas.

Balwen Welsh Mountain

Small, hardy and easy to manage, this breed has a black, brown, or sometimes dark grey coat, with a distinctive blaze of white on the face and four white 'socks'. Renowned for being very resilient health-wise, and for having good feet, which rarely require attention, it needs little supplementary feeding and is an easy-lambing breed.

Hannah Bowen

Berrichon

A large, white-faced sheep from the Cher region of France, the breed has gained as a terminal sire. The triangular-shaped head makes for easy lambing, and lambs are quick to get up on their feet and suckle. They have little or no hair around the genitals, so they are unlikely to need crutching (see Glossary) in summer to prevent flystrike. Wool is of a very high standard and highly prized.

Oogie McGuire

Black Welsh Mountain

Light-boned and easy to manage, the Black Welsh Mountain is a small but incredibly tough breed well-equipped to fend for itself and lamb without assistance. Ewes are normally around 45kg and rams 60–65kg. The short, thick wool – which doesn't need to be dyed – is popular with spinners.

Blackface

Thought to be the most numerous breed in Britain, with more than 1.7 million ewes (11% of the entire British purebred flock), the Blackface is a striking sheep. There are three types – the Perth type, which is a large-framed sheep found mainly in north-east Scotland, south-west England and Northern Ireland; the Lanark type, which is smaller and has shorter wool; and the Northumberland, which is large and soft-woolled and used in breeding the North of England Mule.

Bleu du Maine Sheep Society

Bleu du Maine

A sturdy sheep, with ewes weighing between 80–100kg and rams up to 140kg, it has a dark slate-blue head and legs, and very bright, prominent eyes. Very prolific, the ewe has a wide pelvis, which makes for easier lambing. Often crossed with other breeds to improve conformation and fertility, the pure Bleu du Maine has the advantage of not running too fat when taken to heavier weights.

J. Eveson

Bluefaced Leicester

Characterised by its Roman nose and large, upright ears, the Bluefaced Leicester is the most popular breed in the UK for cross-breeding and rams are used with Swaledale, Blackface, Cheviot, or Welsh Mountain ewes to create the Mule. Mules make up almost half of the crossbred ewes in the UK.

J & K Semple

Border Leicester (R)

A striking animal, due to not only its size but also because of its proud stance. An elegant sheep, it has alert, rabbit-like ears and stands tall and strong. Prized as a pedigree sheep in its own right, the rams are used as terminal sires, but are not as popular as the Bluefaced Leicester. A mature ram can weigh up to 175kg, with ewes smaller, up to 120kg. Categorised as a 'minority' breed by the RBST.

Vicki Beesley

Boreray (R)

The Boreray's status is described as 'endangered' by the RBST. A small primitive breed, mainly found on the remote Scottish island of Boreray, it weighs between 30kg (ewes) and 45kg (rams), making it useful for conservation grazing. Hardy, disease-resistant and able to thrive on poor grazing, it can survive where other sheep would not.

Sara Wood

British Milksheep

Originally known as the Alderbred, this sheep was developed by Lawrence Alderson and is said to be the UK's most prolific breed, with triplets common and a high milk yield. Thought to have been created from a variety of breeds including the Bluefaced Leicester, Dorset, East Friesland and Lleyn, development work began in the 1960s, and the breed was established by the 1980s, but numbers were severely hit in 2001 when foot and mouth disease meant large concentrations had to be culled.

S. Aspital

British Rouge

In its native France, the breed is known as *'Rouge de L'Ouest'* ('red of the west') – a reference to the reddish skin colour and its roots in the Loire region. Although originally kept as a dairy sheep (its rich milk was considered perfect for producing Camembert cheese), selective breeding improved carcass conformation to produce a high meat-to-bone ratio and good muscling.

Cambridge

Developed at Cambridge University in the 1960s by crossing Finnish Landrace rams with breeds including Clun Forest, Llanwenog, Lleyn and Kerry Hill, this medium-sized, dark-faced sheep is regarded as one of the most prolific of all breeds. Fans say it is hardy and has good teeth and low incidence of mastitis.

Castlemilk Moorit Society

Castlemilk Moorit (R)

One of the larger primitive breeds, this was developed early in the 20th century using Manx Loghtan, Moorit Shetland and wild Mouflon sheep by Sir Jock Buchanan-Jardine of Castlemilk Estate in Dumfriesshire. Produces lean, low-fat meat and high-quality wool. Classed as 'vulnerable' by the RBST.

Thomas Lloyd

Charmoise Hill

A powerful-looking, hardy breed from France, which has a deep body with good conformation. It has well-muscled loins and strong gigots, whilst being light-boned to give a good return in meat. Developed in France in the late 18th century, it was the first continental breed to be imported to the UK. Charmoise blood has helped create other popular breeds like the Rouge and the Charollais.

Charollais

With a reputation for being docile, easy to lamb, and for producing fast-growing offspring, the Charollais is popular as a pure breed in its own right and for crossing with others. Lambs are active and quick to suckle and are able to reach 40kg live weight in just eight weeks. Both ewes and rams have long, productive breeding lives.

Cheviot

This white-faced hill sheep from the Scottish Borders can trace its history back to the 14th century. There are South Country, North Country and Brecknock Hill Cheviots. Extremely hardy, with a dense fleece, it is often crossed to produce the Cheviot Mule.

Court-Llacca Cluns

Clun Forest

Originating from the Clun Forest on the Powys/Shropshire border, this brown-faced sheep with a distinctive alert appearance is mainly used in breeding commercial crossbreds. Purebred rams can reach as much as 20kg in 12 to 17 weeks. Ewes have long breeding lives, often still lambing at 12 years old.

Cotswold Sheep Society

Cotswold (R)

Introduced to the UK by the Romans, this large, thick-woolled sheep is sometimes referred to as the Cotswold Lion because of its distinctive 'mane'. Developed in the Cotswold Hills, once an important wool production area, it lost favour with the decline of exports from England. Classed as 'at risk' by the RBST, its numbers are slowly improving.

Dalesbred Sheep Breeders' Association

Dalesbred

Sharing its origins with the Swaledale, this hill breed from Yorkshire is hardy enough to withstand the toughest of weather conditions. The Dalesbred Sheep Breeders' Association was created in 1925 when a group of breeders within the Swaledale Sheep Breeders' Association decided to break away and develop sheep with different characteristics.

Michelle Osbourne

Katherine Walsh

Dartmoor (R)

There are two types, the Whiteface Dartmoor, which is one of the oldest UK breeds and is classed as 'at risk' by the RBST, and the Greyface Dartmoor – also known as the 'improved' Dartmoor – which, although more numerous, is still listed as a 'minority' breed. Both types have long, curly coats, and lambs as well as adults often have to be shorn early to avoid flystrike.

Derbyshire Gritstone

Thrifty sheep that do well with poor grazing on high ground, Derbyshire Gritstones are economical to raise and have the added benefit of producing fine, valuable wool much favoured for hosiery. Sturdy and fast to mature, this is an adaptable breed, which does well in a range of environments.

Chris Birch

Jim Rowe

Devon and Cornwall Longwool (R)

This large, distinctive sheep was produced by crossing the South Devon and the Devon Longwool. Wool from the fleece is hard-wearing and is used in carpets. It's a lean breed, which can be grown on to heavier weights without laying down too much fat. Classed as 'vulnerable' by the RBST.

Devon Closewool (R)

Originating from Exmoor, the Devon Closewool is now seen in Cornwall, Devon, Somerset and parts of Wales. It is, however, classed as a 'minority' breed by the RBST. Its dense fleece helps it cope with strong winds and rain, making it a hardy breed. Docile and easy to manage, it is a medium-sized breed, which is good for beginners.

Nicky Morgan

David & Ruth Wilkins

Dorper

Created in South Africa by crossing Blackhead Persian ewes with Dorset Horn rams, the Dorper arrived in the UK at the start of the century. A naturally shedding sheep with short, fine hair, it requires no shearing or crutching, and it is polyoestrus – able to conceive all year round. The skin is smooth and sought after for leather making. Normally white with a black head and neck, there is also a pure white version.

Dorset Down (R)

Quiet and manageable, this medium-sized, thickset sheep will breed from July onwards to produce fast-growing, early lambs for the spring market. Classed as a 'minority' breed, it produces a short, fine fleece, which is regarded as one of the best in the UK. Lambs have close, tight wool, which protects them from harsh weather.

Dorset Horn (R) and Polled Dorset

Two types exist – the Dorset Horn and the Polled Dorset, with the horned variety less popular and classed as a 'minority' breed. Able to breed out of season, all year round, this strong, adaptable sheep can be reared in any environment to produce lambs that mature early.

Dorset Horn & Polled Dorset Sheep Breeders' Association

159

Easy Care

Designed to make shepherding less demanding, this breed sheds its fleece, removing the need for both shearing and tail docking, lambs without problems and is disease-resistant. The foundation of the breed was the naturally shedding Wiltshire Horn, which was crossed with other breeds including Welsh Mountain. Horns were bred out for easier management.

Morgan Owen

Peter Baber

Exlana

Another low-maintenance hybrid which is gaining support, the sheep gets its name from the Latin word meaning 'without wool'. Developed by a small group of farmers in the south-west of England, the aim was to produce an easy-lambing sheep which naturally shed its fleece, had good worm and foot rot resistance, and gave a good carcass.

K & A Lane

Exmoor Horn

Although a hill breed, this sheep is considered docile and easy to handle and produces a good-quality fleece, which is sought after. The breed is perfect for grazing marginal and species-rich grassland, so plays an important part in conservation grazing. Ewes are often put to a Bluefaced Leicester ram to produce a very marketable mule with good conformation.

Friesian (or Friesland)

Bred primarily as dairy sheep and originating in Friesland, a province in the Netherlands, there are different types of Friesian sheep, including the East Friesian (Friesland) Milk Sheep, the Dutch Friesian Milk Sheep and the Zeeland Milk Sheep. Prolific milkers, they are often crossed with other breeds to produce dual-purpose sheep.

Whitehall Farm

Gotland

From the Swedish island of the same name, this breed was developed by the Vikings using sheep brought back from Russia. It's much prized for its silver-grey, lustrous, curly coat and was originally kept for wool and skins. Ewes lamb easily and can be prolific, with triplets not uncommon. The breed was first imported into the UK in the 1970s and had gained favour for being docile and easy to handle.

Hampshire Down Breeders' Association

Hampshire Down

A native breed, popular in its own right, but also as terminal sire because it produces a well-fleshed carcass with, reputedly, one of the highest 'eye muscle' (see Glossary) scores of any breed. Developed by crossing the Wiltshire Horn and the Berkshire Knot with the Southdown, this is a predominantly white breed with dark brown face and ears.

J & M Cuthbert

Hebridean

Thought to have their roots in Iron Age times, Hebrideans remain hardy sheep, which are able to survive harsh weather and poor grazing. A primitive breed, they are particularly favoured in conservation grazing programmes, partly because they are light-footed, but also because they keep down invasive grasses that threaten other, more important species.

Herdwick

Found in documents going back to the 12th century, the name comes from the Old Norse word, 'Herdwyck', meaning 'sheep pasture'. The hardiest of all UK breeds, it is not unusual to find them grazing at 3,000ft or higher in the Lake District, where most of the population is concentrated. Coat colours range from soft cream through to steel blue, with both rams and ewes having clean, white 'hoar-frosted' legs and faces.

Hill Radnor Flock Book Society

Hill Radnor (R)

With fewer than 900 registered breeding ewes, the Hill Radnor is classed as 'at risk' by the RBST. The sheep is white with a tan face and is larger than most other Welsh hill breeds. Ewes are often crossed with Texels, Charollais, or other terminal sires to produce meatier, faster-finishing lambs. The population is still largely confined to the Radnor/ Brecon areas of mid-Wales.

Jill Tyrer

Icelandic

Domesticated by the Vikings as early as AD 874, Icelandic sheep are thought to be one of the purest breeds in the world. A primitive breed, related to Hebrideans, North Ronaldsays, Shetlands and Soays, they are hardy and easy to tame. This easy-to-manage breed produces meat that is delicately flavoured and considered a gourmet dish in Iceland.

C. Renaud

Ile de France

Developed in France in the 19th century by crossing a Dishley Leicester ram with a Merino ewe, this is a large, muscular, well-proportioned sheep, which can breed out-of-season, allowing three lambings in two years. The breed does not lay down too much fat when kept longer, making it popular for mutton.

Mrs Kershaw

A. Unwin

Jacob

Originally a Mediterranean breed, the Jacob has been in the UK since the 1750s and has become a firm favourite with smallholders because of its attractive fleece. Hardy and adaptable, this breed takes its name from a story in the Bible of how Jacob became a breeder of pied sheep. They can have two or four horns.

Kerry Hill

Distinctive with its black and white markings, this breed originates from the town of Kerry, on the Welsh border. It once suffered a decline in popularity, which gave it a place on the RBST Watchlist, but numbers have now recovered sufficiently to take it off the register.

Lacaune

France's most widely kept dairy sheep, used extensively in the production of Roquefort and other cheeses. The breed is second only to the Friesian in popularity in Europe and the US and, although producing less milk, is regarded to give a better-quality product with a higher fat content. The Lacaune tends to shed its wool from the chest down, removing much of the need for shearing.

Chris Sander

Leicester Longwool (R)

Large but surprisingly docile, this breed is classed as 'vulnerable' by the RBST. Its long, lustrous fleece is used for spinning and there are two permitted colours: white and black. Favoured for producing large but lean-finished lambs, rams are often used as terminal sires.

Lincoln Longwool (R)

The largest of the UK's native breeds, with ewes weighing up to 110kg and rams up to 160kg, the Lincoln is in the RBST's 'at risk' category. Like the Leicester Longwool, it is easy to manage. There are both white and black strains and the fleece produced is one of the heaviest of any breed and is popular with hand-spinners.

Sion Morgan

Julie Hyett

Llandovery Whiteface Hill

Bred to withstand the harsh upland weather of south-west Wales, these sheep are strongly built and have a dense fleece. Ewes are often crossed with lowland and longwool breeds to produce Welsh Halfbred and Mule ewe lambs.

Llanwenog (R)

This west Walian breed was developed by crossing with the Shropshire Down and is a medium-sized, docile sheep, which lambs easily and has a reputation for being prolific and easy to lamb, and with a high-quality fleece. Numbers have dropped, however, and in 2015 it was classified a 'minority' breed by the RBST.

Lleyn Sheep Society

Lleyn

From the Llŷn Peninsula in Gwynedd, north Wales, this is a breed renowned for its mothering ability and adaptability to different environments. Medium-sized and easy to manage, they are good for beginners and lamb easily outside or in. They are very popular for crossing with Suffolk, Charollais and Texels.

Jane Cooper

Lonk

Developed in Lancashire and the Yorkshire Pennines, the Lonk is hardy, strong-boned, and long-lived. A medium-sized sheep with a good fleece and a lean carcass, it's thought the name may have been derived from the word 'lanky', meaning tall and thin.

Cathy Cassie

Manx Loaghtan (R)

A small primitive sheep, which originates from the prehistoric short-tailed breeds of sheep found in isolated parts of north-west Europe. It has magnificent horns – either two or four are seen in both ewes and rams. Some will shed their fleeces naturally from about May, but most will need shearing. The RBST classes the Manx Loaghtan as a breed 'at risk'.

Edward Faulding

Masham

A small market town in North Yorkshire gives this tough breed its name. The Masham is the result of crossing a Teeswater ram with ewes such as Dalesbred, Swaledale, or Rough Fell. Medium-sized and polled, they are economical to rear, fattening well even on rough grazing, and doing even better in less challenging environments.

Meatlinc

A specialist terminal sire, this hybrid was created more than 40 years ago specifically to provide rams for commercial flocks to produce fast-growing, muscled lambs. The breed is managed by a commercial organisation, the Meatlinc Sheep Co. Ltd, and rams are sold as shearlings, ready to breed.

Merino

Originally from south Portugal, the Merino gained worldwide popularity for its fine, soft wool. Although primarily a medium-sized sheep raised for wool, some strains have been developed more for meat and have inferior fleeces. Excellent at foraging, the breed is very adaptable to different environments.

Mouflon

Considered to be one of the ancestors of all modern day sheep (see Chapter 1), the rough-coated Mouflon originated on the islands of Corsica, Sardina and Cyprus. In the wild, rams and ewes live in separate groups apart from during the mating season. They have a strict hierarchy: rams don't mate until they are several years old because they need to work their way up the pecking order.

Suzannah Coke & Kerry Gibb

Norfolk Horn (R)

Believed to be descended from the ancient Saxon black-faced sheep of northern Europe, this is a breed considered 'at risk' by the RBST. Excellent foragers, they are able to thrive on rough grazing without losing condition, and are often used in conservation grazing. Medium-sized, their narrow heads and shoulders make for easy lambing.

Jane Cooper

North of England Mule

Originating from upland and hill farms in northern England, this medium-sized, crossbred sheep is popular across the UK. The breed is created by crossing a Bluefaced Leicester ram with a Swaledale or a Northumberland-type Blackface – the latter being used to pass on hardiness and the ability to graze rough land.

J. R. Lane

North Ronaldsay (R)

A small primitive breed thought to date back to the Bronze Age, the North Ronaldsay, from the most northerly of the Orkney Islands, is classed as 'vulnerable' by the RBST. Rams are horned, but ewes can be either horned or polled. The fine fleece can come in any colour – white, black, or various shades of grey and brown. Because of the location of the breed, it developed the ability to thrive on a diet of seaweed.

Tina Archer

Ouessant

Believed to be the world's smallest natural breed, the Ouessant ranges from just 45 to 50cm tall (about 5cm shorter than the smallest native British breed, the Soay) and weighs a maximum of 20kg. It comes from an island off the west coast of Brittany, which gives it its name, and is a docile and easy-to-handle breed often used for grazing orchards.

Oxford Down Sheep Breeders' Association

Oxford Down (R)

Created by crossing Cotswold rams with Hampshire Down and Southdown ewes, Oxfords are tough and productive. The heaviest of all the Down breeds, the Oxford is fast-maturing and popular for producing early-season lamb. Its heavy fleece is commercially valuable and is used in hosiery, knitting yarns and felts. Classed as a 'minority' breed by the RBST.

Portland Sheep Breeders' Group

Portland (R)

A small, primitive heathland breed from Dorset, the breed is one of only a few that are capable of breeding at any time of the year, and normally has just one lamb. Both males and females are horned and have short, creamy fleeces with tan faces and legs. Lambs are born a deep red and lighten with age. Classed as 'at risk' by the RBST.

Tina Archer

Romney

Formerly known as the Romney Marsh or the Kent, the Romney is a longwool breed with either a white or a coloured fleece, and which is used for a wide range of products. Although a large sheep, it's fairly easy to manage. Neither ewes nor rams have horns. Some flocks are grazed on salt marshes in Kent to produce Romney Salt Marsh lamb.

Rough Fell Breeders' Association

Rough Fell

A strong, hardy native breed of Cumbria, this is a medium-sized mountain sheep, which does well on poor upland grazing. The thick fleece – which is largely exported for carpet making – provides excellent protection against the worst of weathers.

Andrea Molyneux

Roussin

Breeds from the coastal areas of northern France were crossed with Dishley Leicester, Southdown and Suffolk to create the hardy Roussin. Rams will work out of season, making them suitable for early lambing flocks. A medium-sized breed, its coastal origins make it suitable for challenging climates.

Ryeland

Placid and manageable and sought after by smallholders, there are two distinct types – the Ryeland (left), which is pure white, and the Coloured Ryeland (below), which can be silver, brown, fawn, or black. A recessive gene in the breed means that sometimes white parents produce coloured lambs. The name is a reminder that the sheep was raised on rye-growing land in Herefordshire.

Ruth Spires

Chris Smyth

Eilidh MacPherson

Shetland Sheep Society

Scotch Mule

A cross between a Bluefaced Leicester ram and a Blackface ewe, the Scotch Mule is tough, produces multiple lambs, and is considered an easily handled breed. Inheriting survival traits from the hill breed in its ancestry, it will cope with most weathers.

Shetland

Popular with small-scale sheep keepers and used widely in conservation grazing, the Shetland is a primitive breed from the Shetland Islands – Britain's most northerly outpost. They do well on rough grassland and scrub and rarely need supplementary feeding. Their fleeces come in 11 main colour types and are in demand for tanning and spinning.

Shropshire Sheep Breeders' Association

Susan Constable-Maxwell

Shropshire

Thought to have originated in the Staffordshire and Shropshire border areas, the Shropshire is a favourite for grazing orchards and Christmas tree plantations because it tends not to damage trees by stripping bark in the way most other breeds do. It's a medium-sized sheep with a heavy, dense fleece, which is used for hosiery and knitting yarns.

Soay (R)

Originating from the island of Soay off the west coast of Scotland, this small and agile sheep has a sought-after fleece and its meat is developing a following in top restaurants. Fine-boned and light-footed, primitive Soays are often used in conservation grazing. They are listed as 'at risk' by the RBST.

Rachel Graham

Southdown

Popular across the world, particularly as a terminal sire, the Southdown, as the name suggests, came from the South Downs of England. A polled breed with a good temperament, it's medium-sized, considered easy to handle and produces fast-growing lambs. A craze for producing miniature Southdowns (pictured above, right) – known as 'Babydoll' sheep – began in the United States and they now have their own breed society there.

Suffolk Sheep Society

Suffolk

Big, powerful and productive, the Suffolk is one of the most popular terminal sires in the UK. Developed by crossing a Southdown ram with Norfolk Horn ewes, it is the largest commercial breed in the UK and can be grown on to give heavier carcass weights without getting too fat. The ewe's wide pelvis avoids lambing problems and the breed is also known to have good feet.

Kay Hutchinson

Piple Cornu

Swaledale

Another hardy breed from the north of England, the Swaledale is popular for crossing with Bluefaced Leicester rams to create the North of England Mule. A medium-sized breed with horns on both sexes, it thrives on poor grazing and lambs easily in all weathers.

Teeswater (R)

A large, longwool breed from the Teesdale area of County Durham, the rams are used as terminal sires for use on a range of breeds. A specialist wool-producing sheep, with fleeces in great demand for wool for clothing, it's a lean breed with a well-developed back end, making it a good meat sheep, too. It's classed as 'vulnerable' by the RBST.

Texel

The Dutch Texel – along with the Beltex and the Charollais – is one of the top terminal sires worldwide. The name comes from the island off the north coast of Holland where it originated. Believed to date back to Roman times, it was a hardy sheep suited to grazing sparsely vegetated sandy soil. The modern Texel is a medium-sized breed and highly adaptable to different environments.

Urial

Similar in appearance and behaviour to the Mouflon – but a completely different sub-species of the wild sheep family *Ovis orientalis* – the Urial is another ancient ancestor of our native breeds. Found mainly in western central Asia, the Urial's conservation status has been threatened by loss of habitat due to development for housing and industry.

Emma Collison

Valais Blacknose (Walliser Schwarznasenschaf)

Gaining popularity for its novel appearance, this sheep is from the mountainous regions of Switzerland and so tough and able to withstand harsh weather. A large breed, the wool is used mainly in carpet making and felting. One of the first UK importers spent £55,000 on just ten ewes and one ram.

T. & S. Butcher

Vendéen

From the west coast of France, in the Vendée region, this is a medium-sized breed, which will lamb naturally out of season without need for sponging. Lambs are fast growers, and ewes can have three lots of lambs in the space of two years. The breed is bred more for meat than the quality of its wool.

Paul Johnson

Welsh Halfbred

A small and manageable breed, which is a cross between a Welsh ewe and a Border Leicester ram. Able to withstand high altitudes and poor grazing, they are good foragers and therefore economical to keep, and can be stocked at higher densities than some other breeds.

Welsh Hill Speckled Face Sheep Society

Welsh Hill Speckled Face

Very hardy and adaptable and easy to lamb, this is a small- to medium-sized sheep, which has a good temperament and is easy to manage. Similar in appearance to the Kerry Hill, the ewes are crossed with a Bluefaced Leicester ram to produce the much-sought-after Welsh Mule. The sheep has distinctive markings, with a white face and black markings around the eyes and ears. Another speckled-faced welsh sheep is the Eppynt and Beulah Speckled Face, which is popular for crossing with continental and lowland rams.

Welsh Mountain Sheep Society

Welsh Mountain

Economical to keep, resilient and easy lambing, Welsh Mountain Sheep are seen across the UK in upland and lowland systems. Ewes are often crossed with a Bluefaced Leicester ram to produce the Welsh Mule, or with a Border Leicester for Welsh Halfbreds. There are regional variations – in south Wales there are two types, the Nelson and the Glamorgan, which are both medium-sized sheep; the North Wales type is smaller but considered hardier.

Llŷr Jones

Welsh Mule

Robust Welsh Mules are the result of crossing a Bluefaced Leicester ram with Welsh Mountain, Beulah, or Welsh Hill Speckled Face ewes. Prolific and easy lambers and adaptable to indoor or outdoor systems, they produce fine, good-quality wool.

Freda Magill, Redbridge Studio

Wensleydale (R)

Known for its impressive fleece, this sheep is classed as 'at risk' by the RBST. A large sheep, with rams sometimes exceeding 140kg, it's often crossed with hill ewes to pass on its strong carcass and wool-producing traits. Wensleydale wool is regarded as the finest and most valuable lustre longwool in the world.

Whitefaced Woodland (R)

A large breed from the South Pennines, sometimes known as the Penistone sheep, this is one of the largest of the British Hill breeds and is strong-framed and powerful with distinctive spiralling horns. A good meat breed, it also produces wool finer than that of many other hill breeds, and so suited to knitting or hosiery. It's listed as 'vulnerable' by the RBST.

Wiltshire Horn

This breed has a short fleece, which it sheds naturally in early spring, making it perfect for those who dislike shearing and making it less prone to flystrike. A large-framed sheep, it's used as a terminal sire to produce lambs that mature quickly and which often inherit the fleece-shedding gene.

Zwartbles Sheep Breeders' Association

Zwartbles

Thought to trace its ancestry back to a large breed called the Schoonebeker in north-east Holland, Zwartbles are tall, elegant and well-built sheep, which are naturally polled. The breed has a distinctive look, with a solid black or chocolate brown base colour and a clear white blaze on the face.

Appendix 2

THE SHEPHERD'S CALENDAR

You will soon establish your own calendar of routine flock management and husbandry tasks, but here is a basic guide to what you should be thinking about doing and when. This calendar is based on the traditional – and easy to remember – system of putting the rams in on Bonfire Night (5 November) to achieve a lambing season starting around April Fools' Day (1 April). With earlier or later lambing systems, adjust accordingly.

There is a really useful resource on the Eblex website (www.eblex.org.uk), which will help you create your own management calendar, complete with recommended tasks built in. You can also add your own information, too. All you need to do is enter the date your ram is turned in or the date you would like to start lambing. You can sign up to receive email reminders of when particular tasks are due, which can be a help at busy times.

In all the old sheep textbooks, the shepherd's calendar starts in September, before tupping. In spring-lambing flocks, the preparation work for next year's lambing season normally starts then. There is also a traditional, very practical reason as to why everything is deemed to start in September – because, historically, lots of farm tenancies used to begin that month – to allow crops to be harvested by a farmer giving up the tenancy and to allow the new tenant to plant for next year.

However, as you may be starting with sheep at any time of the year, this particular calendar is designed to fit in with the average person's diary or kitchen wall calendar.

MONTH	TASKS
January, February, March	Keep an eye on condition of pregnant ewes. Scan ewes in January, group according to lambs expected and feed accordingly. If you vaccinate for clostridial diseases and pasturella, give ewes a booster four to six weeks before the due date. Allow time to treat ewes not previously vaccinated, as they will need two doses, four to six weeks apart. If you've had a problem with orf in the past, discuss prevention methods with your vet. Prepare lambing shed if lambing indoors or set up outdoor pens and 'nursery' area, ideally with shelter. Check your lambing kit and replace any missing items (see Chapter 10). Get yourself on a lambing course or arrange some practical experience with someone who is lambing earlier than you.
April	Lambing begins! Refer to Chapter 10 for reminders on what to look out for. Record ewes that abort or have delivery problems and consider future culling. If lambing outdoors, predator control may be important. Castrate lambs and dock tails if part of your management routine. If the weather is mild, protect against flystrike (Chapter 6). Check the length of coverage on insecticides and make a note of when to repeat applications. Discuss faecal egg counts and worming plans with your vet. Check the SCOPS and NADIS websites for regional updates and news alerts. Talk to the vet about protection against coccidosis and order vaccines for lambs for clostridial diseases/pasturella.
May, June	If you've been feeding concentrates to the ewes, start gradually decreasing as soon as the grass starts to grow well. Be on your guard against flystrike and watch for warning signs. Start 'crutching' (see Chapter 6) in May and plan shearing time. Give lambs a second dose of vaccine against clostridial diseases/pasturella.
July	Start weaning lambs on to fresh grazing. Continue to watch for signs of worms in both adult sheep and lambs. Arrange faecal egg counts and treat accordingly. Continue to watch for flystrike. Repeat pour-on applications if necessary (check instructions for frequency of applications).
August	Start thinking about the next breeding season. Condition-score the breeding flock. Give rams an 'MOT' (see Chapter 9). Get ewes blood-tested for possible deficiencies and discuss supplements and abortion vaccines with the vet.
September	Sell or cull under-performing stock. Source replacements/new rams at annual sales or from private breeders. Discuss fluke risk and protection with your vet. Weigh ewe lambs with a view to future mating. Depending on breed and grazing quality, some spring lambs may be ready for slaughter.
October	Flush underweight ewes prior to mating. Check feet and carry out any internal parasite treatments ahead of mating. If 'teasers' are used, put them in mid-month.
November	Rams go in on the 5th. Raddle and watch for repeat matings. Provide plenty of hay/haylage to supplement grazing. Cover feeders to keep hay dry if possible, and always watch for mould.
December	Remove rams 17 December.

GLOSSARY

Abomasum – the fourth compartment of the stomach of a ruminant.

Abortion – the premature loss of a pregnancy.

Afterbirth – the placenta and foetal membranes expelled from the uterus after lambing.

Amino acid – one of the building blocks of protein.

Anaemia – a lower than normal number of red blood cells.

Anthelmintic – a drug that kills certain types of intestinal worms.

Antibiotic – a drug that inhibits the growth of/destroys microorganisms.

Antibodies – proteins produced by the immune system to fight specific bacteria, viruses, or other antigens.

Antitoxin – an antibody that can neutralise a specific toxin.

ARAMS – Animal Reporting and Movement Service, which is the online recording system for movements of sheep, goats and deer in the UK.

Artificial Insemination (AI) – placing semen into the uterus by artificial means.

Banding – the process of applying rubber bands to the tail/scrotum for docking/castration.

Bloat – excessive accumulation of gases in the rumen.

Bolus – an object placed in the reticulum, containing slow-release medication or electronic identification information.

Bottle jaw – oedema or fluid accumulation under the jaw. Often a sign of infection with haemonchosis.

Breech – delivery position in which the lamb is presented backwards with its rear legs tucked underneath.

Broker/broken-mouthed – a sheep that has lost or broken some of its incisor teeth, usually due to age.

Bummer – in the United States, a slang term for an orphan lamb.

Burdizzo – bloodless castration method, which involves crushing the blood vessels leading into the testicles.

Cade lamb – an orphan lamb, or one rejected by its mother.

Caprinae – the group of even-toed ungulates, which includes sheep and goats.

Carding – process of using a metal-pronged carding comb or brush to either groom a sheep for showing or prepare a fleece for spinning.

Cast – used to describe a sheep on its back, unable to regain footing.

Castrate – process of removing the testicles (verb); a male sheep which has had the testicles removed (noun).

Cellulose – a component of plant cell walls that is not digestible by most animals.

Cervix – the lower section of the uterus that protrudes into the vagina and dilates during labour to allow birth.

Clostridial diseases – potentially fatal infections caused by clostridia bacteria.

Cryptorchidism – failure of one or both testes to descend.

Coccidiostat – any of a group of chemical agents mixed in feed or drinking water to control coccidiosis in animals.

Colostrum – first milk a ewe gives after birth. Rich in antibodies, it helps protect newborns against disease.

Combing – straightening of wool fibres and removal of short fibres and other impurities.

Concentrate – high-energy, low-fibre feed that is highly digestible.

Conception – fertilisation of an egg by a sperm.

Corpus luteum – a yellow, progesterone-secreting mass of cells formed after a mature egg has been released from an ovarian follicle.

CPH number – County Parish Holding number, which registers land as being for agricultural use.

Crimp – the natural waviness of the wool fibre.

Cull ewe – a ewe no longer suitable for breeding, which is sold for meat.

Culling – slaughtering an unwanted animal.

Creep feed – small pellets of high-protein supplementary feed given to lambs.

Creep grazing – allowing young animals to forage in areas with restricted access, while older sheep are kept out.

Cross-grazing – using two or more species of animals on the same land, because they graze in different ways and benefit the sward.

Crutching (see 'Dagging')

Cud – food that is regurgitated by a ruminant to be chewed again.

Dagging – also known as 'crutching', the removal of wool from around the tail and between the rear legs of a sheep.

Dags – wool contaminated with faeces on the back end of a sheep.

Dam – mother.

DE – digestible energy.

Diarrhoea – also known as 'scouring'; an unusually loose or fast faecal excretion.

Docking – shortening the tail to avoid build-up of faeces and so reduce the risk of flystrike. Also known as 'banding'.

Draft or draught ewe – a ewe too old for rough hill grazing, which is drafted out of the flock and moved to another farm to better grazing.

Drench – an orally administered liquid medicine (noun); to administer a liquid medicine (verb).

Dressing percentage – the percentage of the live animal that ends up as carcass.

Droving/driving – walking sheep from one location to another.

Dry ewe – one that is barren or not currently carrying a lamb.

Dystocia – difficulty in giving birth or being born.

EID tag – a yellow electronic identification tag, required for ID purposes in the UK.

Elastrator – a scissors-like tool used to apply tight rubber bands to the tail and scrotum for docking and castration purposes.

Embryo – an animal in the early stage of development before birth.

Embryo transfer – implantation of embryos or fertilised eggs into a surrogate mother.

EBV – Estimated Breeding Value. A prediction of breeding potential eg. quality and expected growth rate of lambs produced.

EPD (expected progeny difference) – the predicted difference between the performance of an animal's progeny and the average progeny performance of all the animals in the breed.

Epididymis – tiny tube where sperm collect after leaving the testis.

EUROP – a system of carcass classification, where E = excellent, U = Very good, O - fair and P = poor.

Ewe – adult female sheep.

Eye muscle – the muscle running the length of the loin; the round or oval-shaped piece of lean meat in a loin chop.

Faecal egg count (FEC) – the process of assessing the number of worm eggs in a given measurement of faeces.

Fertiliser – natural or synthetic soil improvers, which are spread over or worked in to improve fertility.

Flock number – needed in the UK for identifying animals belonging to an individual herd.

Flushing – increasing nutrition in the few weeks before mating to improve fertility, or in the period before birth to increase lamb birth weight; sometimes referred to as 'steaming up'.

Flushing (eggs/embryo) – removing a fertilised or unfertilised egg from an animal as part of an embryo transfer procedure.

Fodder crop – a plant grown for animal feed.

Footbath – a long trough filled with a chemical preparation, which sheep stand in for protection from/treatment of hoof conditions like scald and foot rot.

Foot rot – infectious pododermatitis, a painful, bacterial infection affecting sheep, goats and cattle.

Forage – edible plant material used as livestock feed.

Fostering – encouraging a ewe to accept a lamb from another in the flock.

Gestation – the length of pregnancy. In most sheep breeds, approximately 145 days.

Gimmer – a female sheep, normally over a year old and before her first lamb.

Greasy wool – wool as it comes off the sheep, unwashed and still containing lanolin.

Halal – a set of Islamic dietary laws, which regulate the preparation of food.

Half-bred – a 'mule' or cross-bred sheep, e.g. a mountain ewe crossed with a longwool ram.

Heat – the period when a ewe is fertile and receptive to the ram; also known as 'oestrus'.

Hectare – a metric unit of area equal to 10,000 square metres, or 2.471 acres.

Hefting (or heafing) – in certain breeds, a natural instinct to keep to a particular area or 'heft' throughout life.

Heterosis – the increase in performance (e.g. fecundity, yield, growth rate) of hybrids compared to that of their purebred parents.

Heritability – the extent to which a trait is influenced by genetic make-up.

Hog, hogg, or hogget – a young sheep of either sex aged between one year and two years, or until its first teeth erupt. The meat is also described as 'hogget'. 'Teg' is another name for a sheep of this age.

Hybrid vigour – an increase in performance due to cross-breeding.

Hyperthermia – Elevated body temperature; overheating.

Hypothermia – lower than normal body temperature.

Immunity – a natural or acquired resistance to specific diseases.

Inbreeding – the mating or crossing of closely related animals; sometimes referred to as 'line-breeding' when carried out to pass on or strengthen certain desirable traits.

In lamb – pregnant.

Intramuscular (IM) injection – one that is given straight into a muscle.

Intravenous (IV) injection – one that is given directly into a vein.

Joint ill (infectious polyarthritis; navel ill; pyosepticaemia) – in newborn lambs, an inflammation of the joints caused by bacteria entering the body, normally via an untreated umbilical cord.

Jugular – veins in the neck that carry deoxygenated blood from the head back to the heart.

Ketone – an acidic substance produced when the body uses fat instead of sugar for energy.

Ketosis – a metabolic disorder where ketones build up in the body.

Kosher – food prepared in accordance with Jewish dietary laws.

Lactation – the production and secretion of milk; the period when the ewe produces milk.

Lamb – a young sheep (noun). To give birth to a lamb (verb). The meat from an animal less than one year old (noun).

Lambing percentage – the number of lambs (including multiple births) successfully reared in a flock compared with the number of ewes that have been mated.

Lanolin – a thick yellowish grease secreted by the sheep's skin; variously known as 'wool wax', 'wool fat', 'wool grease', or 'yolk'.

Line-breeding (see also 'inbreeding') – the mating of closely related animals within a particular blood line; inbreeding.

Live vaccine – a vaccine in which a live virus is weakened through chemical or physical processes to produce an immune response without causing the effects of the disease.

Lutenizing hormone (LH) – the hormone that triggers ovulation and stimulates the *corpus luteum* to secrete progesterone; in males, it stimulates testosterone production.

Mastitis – an uncomfortable inflammation of the mammary glands due to bacterial infection.

Micron – measurement unit for wool-fibre diameter; one millionth of a metre.

Moiled – hornless; polled.

Monorchid – a male with only one testis descended into the scrotum.

Morphology – the size and shape of sperm.

Motility – the ability of sperm to move.

Mule – a cross-bred sheep produced by mating an upland/hill ewe with a lowland ram (normally a Bluefaced Leicester).

Mutton – the meat from a sheep aged two years or older.

Necropsy – a post-mortem examination.

Nematode – a parasitic roundworm.

Oestrogen – female sex hormone produced by the ovaries, which is responsible for the oestrus cycle.

Oestrus – the 'heat' period, during which ewes are fertile and receptive to the ram.

Oestrus cycle – the reproductive cycle of the female.

Omasum – the third part of the ruminant stomach, between the reticulum and the abomasum.

Orf – a virus that causes contagious ecthyma in sheep and goats; a zoonotic infection – i.e. can be passed to humans.

Ova/ovum – female sex egg, also known as an oocyte.

Ovine – relating to or resembling a sheep.

Ovis – a taxonomic genus within the sub-family 'caprinae'.

Ovulation – the release of mature eggs from the ovary.

Oxytocin – a naturally secreted hormone that encourages the contraction of the uterine muscles during labour and milk let-down; a veterinary product used to stimulate contractions and help lactation.

Parturition – the birthing process.

Pedigree – a 'family tree' showing the ancestry of a registered, purebred animal.

Pelt – the sheep's skin, complete with wool.

Pet lamb – an orphan lamb which is artificially reared.

pH – a value that indicates the acidity or alkalinity of something (e.g. rumen, soil).

Phenotype – the observable physical characteristics of an individual.

Pink eye – infectious keratoconjunctivitis; an eye condition in which the conjunctiva become inflamed or infected.

Pizzle – penis.

Placenta – the organ that protects and nourishes the foetus(es) while in the uterus.

Pneumonia – an inflammation of the lungs, caused by a bacterial or viral infection.

Poddy lambs – orphan lambs.

Polled – without horns.

Polyoestrus – able to breed all year round.

Pour-on – a chemical preparation for control of internal/external parasites, which is applied to the fleece of the sheep and is gradually absorbed; an alternative to injectable treatments.

Predator – an animal that lives by hunting, killing and eating other species.

Probiotic – a living organism used to manipulate fermentation in the rumen.

Progeny – the offspring of an animal.

Progesterone – a female hormone produced in large quantities by the placenta during pregnancy and secreted by the ovaries.

Prolific – highly productive in lambing terms; fecund.

Purebred – not crossed with another breed.

Quarter – half of the ewe's udder.

Raddle – coloured pigment applied to the ram's brisket to mark the females he mates; a harness used to hold a raddle crayon.

Ram – an uncastrated adult male sheep; a tup.

Reticulum – the second chamber of the ruminant digestive tract.

Ringwomb – failure of the cervix to dilate sufficiently, which causes delivery problems.

Rotational grazing – organised system of moving stock from one grazing unit to another.

Roughage – high-fibre feed, which is low in both digestible nutrients and energy (e.g., hay, straw, silage).

Roundworm – unsegmented parasitic worms with elongated rounded bodies, which are pointed at both ends.

Rumen – the first compartment of the stomach of a ruminant animal. It contains bacteria and protozoa, which break down cellulose.

Ruminant – an animal with a multiple-chambered stomach, which is able to digest cellulose.

Scab – an irritating skin condition caused by parasitic mange mite, *Psoroptes ovis*.

Scouring – see 'diarrhoea'.

Scrapie – a fatal, degenerative disease affecting the central nervous system of sheep and goats.

Scrotum – the pouch of skin containing the male's testicles.

Semen – a combination of sperm, seminal fluid and other male reproductive secretions.

Sharps – needles, syringes, scalpel blades and anything else that can puncture the skin.

Shearing – removing the fleece using mechanical clippers or hand shears.

Shearling – a sheep that has been shorn once, normally over a year old; also known as a 'one-shear' or 'yearling'.

Silage – fodder prepared by storing and fermenting grass or other forage plants in wrapped bales or in a silo.

Sire – father.

Skirting – removing the stained, unusable, or less-desirable parts of a fleece.

Spinning – working natural fibres into thread or wool.

Staple – the length of a lock of shorn wool; the longer length wools within a grade.

Stillborn – newborn lamb that is delivered dead.

Stocking density – the relationship between the number of animals and an area of land.

Store – a weaned lamb not ready for slaughter, which is kept for fattening.

Stratification – system in the UK whereby breeds are classified according to their natural environments.

Straw – the stems of cereals like wheat, barley, or oats, which are cut and baled and used for fodder or bedding.

Strip-grazing – controlling grazing by confining animals to specific areas of land (often using electric fencing) for short periods of time before moving them on to fresh ground.

Stun – to render unconscious, e.g. prior to slaughter.

Subcutaneous injection – given under the skin, but not into the muscle; sometimes shortened to 'sub-Q' or 'SQ').

Sustainable farming – an approach that uses on-farm resources efficiently, reduces demands on the environment and may help rural communities.

Tack sheep – sheep overwintered on a different farm with better grass, often moved from an upland to a lowland area.

Tapeworm – ribbon-like parasitic flatworms found in the intestines.

Teaser – a ram that has been vasectomised to prevent reproduction, but which is used to stimulate ewes for mating.

Teg – see hog, hogg, hogget.

Terminal sire – a ram used in cross-breeding to pass on desirable traits to its offspring – most often to improve the carcass..

Testosterone – a hormone that promotes the development and maintenance of male sex characteristics.

Total Digestible Nutrients (TDN) – a system used for expressing the energy value of feeds.

Tup – an uncastrated adult male sheep.

Turnover crate – a device used to restrain and then invert a sheep, e.g. to treat foot problems.

Twin lamb disease – pregnancy toxaemia, a metabolic disease affecting very underweight or overweight ewes carrying multiple lambs.

Udder – the milk-secreting organ.

Ultrasound – a procedure in which high-energy sound waves are used to create images of organs and structures in the body, e.g. in pregnancy diagnosis.

Urea – the main end product of protein metabolism in animals.

Uterus – the organ in which the foetuses develop; the womb.

Vaccine – an injection given to improve resistance to/prevent disease.

Vagina – the passageway from the cervix to the external organs.

Weaning – the process of taking young animals away from their source of milk.

Wet adoption – covering a lamb that is to be adopted with birthing fluids from the ewe's own lamb.

Wether – a castrated male sheep.

Withdrawal period – after treatment with a medical product, the amount of time that must be allowed before meat or milk is allowed into the human food chain.

Yearling – an animal between 1 and 2 years of age.

Yow – in some parts of the north of England, a name for an adult female sheep.

Zero grazing – a system of growing fodder but not allowing livestock to graze it directly; instead, the crop is cut and taken to the animals.

Zoonosis – a disease or ailment which is zoonotic, i.e. one which normally exists in animals, but can be passed to humans.

USEFUL CONTACTS

Government organisations

England: Department for Environment, Food and Rural Affairs (DEFRA)
Tel 08459 33 55 77
www.defra.gov.uk
Rural Payments Agency
Tel. 0845 603 7777 www.rpa.gov.uk

Scotland: Scottish Government
08457 741 741 or 0131 556 8400
www.scotland.gov.uk
Scottish Rural Payment and Inspections Directorate (SGRPID)
Tel 0845 601 7597

Wales: Welsh Government
Tel 0300 0603300 or 0845 010 3300
www.wales.gov.uk

Northern Ireland: Northern Ireland Executive Tel 028 9052 8400
www.northernireland.gov.uk
Department of Agriculture and Rural Development Tel 0300 200 7852
www.dardni.gov.uk

Republic of Ireland: Irish Government
Tel. 0761 07 4000 www.gov.ie
Department of Agriculture, Food and the Marine Tel. 01 607 2000
www.agriculture.gov.ie

Agriculture and Horticulture Development Board (AHDB). Formerly EBLEX.
www.beefandlamb.ahdb.org.uk
Tel. 024 7669 2051
Animal Health Disease Outbreak Information Line
Tel 0844 8844600
Animal and Plant Health Agency
Tel 08459 33 55 77 (Use postcode search facility on website for local contacts in England and Wales)
www.defra.gov.uk/ahvla
Scotland Tel 08457 741 741 or 0131 5568400
Animal Recording and Movement Service (ARAMS) Tel. 0844 573 0137
www.arams.co.uk

British Sheep Dairying Association
www.sheepdairying.com
Tel. 01825 791636
British Veterinary Association
Tel. 020 7636 6541
www.bva.co.uk
British Wool Marketing Board
Tel: 01274 688666
www.britishwool.org.uk
Compendium of Animal Health and Welfare in Organic Farming
www.organicvet.co.uk
Farm Animal Health
www.farmanimalhealth.co.uk
Tel. 01276 694402
Food Standards Agency
Tel. 020 7276 8829 www.food.gov.uk
Scotland Tel 01224 285100;
Wales Tel 02920 678999
Farming and Wildlife Advisory Group (FWAG) Tel. 02476 696699
www.fwag.org.uk
Hybu Cig Cymru/Meat Promotion Wales
www. hccmpw.org.uk
Tel. 01970 625050
Humane Slaughter Association
Tel. 01582 831919 www.hsa.org.uk
Livestock and Meat Commission Northern Ireland (LMCNI) Tel. 028 9263 3000
www.lmcni.com
Mutton Renaissance
Tel. 01684 899255
www.muttonrenaissance.org.uk
National Farmers' Retail & Markets Association (FARMA)
12 Southgate Street, Winchester, Hampshire, SO23 9EF
Tel. 0845 4588420 www.farma.org.uk
National Disease Information Service
www.nadis.org.uk
Tel 07795 557619
National Fallen Stock Company
Tel. 01335 320014 www.nfsco.co.uk
National Office of Animal Health (NOAH)
Tel. 020 8367 3131 www.noah.co.uk
National Sheep Association (NSA)
Tel. 01684 892661
www.nationalsheep.org.uk
Premium Sheep and Goat Health Scheme
www.sruc.ac.uk Tel. 01463 226995

Quality Meat Scotland Tel. 0131 472 4040
www.qmscotland.co.uk
Rare Breeds Survival Trust (RBST)
Tel.02476 696551 www.rbst.org.uk
Scotland EID
Tel. 01466 794323
www.scoteid.com
Small Shepherds' Club
Tel. 01483 546045
www.smallshepherdsclub.org.uk
Soil Association Tel. 0117 314 5000
www.soilassociation.org
Sustainable Control of Parasites in Sheep (SCOPS) Tel. 01684 892 661
www.scops.org.uk

Books

An Introduction to Keeping Sheep (Good Life Press, 2007) Upton, Jane, and Soden, Denis
Artisan Cheese Making (Ten Speed Press, 2011) Carlin, Mary
A Manual of Lambing Techniques (Crowood Press, 2003) Winter, Agnes and Hill, Cicely
Charcuterie: The Craft of Salting, Smoking, and Curing (W.W. Norton 2013) Ruhlman, Michael
Counting Sheep (Profile, 2014) Walling, Philip
Collie Psychology (First Stone, 2013) Price, Carol
Cured (Jacqui Small, 2010) Wildsmith, Lindsey
Home Tanning and Leather Making (Wylie Press, 2008) Farnham, Albert
Field Guide to Fleece (Storey, 2013) Robson, Deborah
Fifteen Grades of Hay (CreateSpace, 2012) Weathersheep, Derek
TheLambs for the Freezer (Crowood, 2011) Kendrick, Sue
Lameness in Sheep (Crowood Press, 2004) Winter, Agnes
Making Your Own Cheese (How To Books, 2010) Peacock, Paul
Mastering Artisan Cheese Making (Chelsea Green, 2013) Caldwell, Gianaclis

Much Ado About Mutton (Merlin Unwin, 2014) Kennard, Bob

Practical Lambing and Lamb Care (Blackwell, 2004) Eales, A, Small, J, and Macaldowie, C.

Sheep (O Books, 2006) Butler, Alan

Sheep Ailments – Recognition and Treatment (Crowood, 2001), Straiton, Eddie

Sheep Health, Husbandry and Disease: A Photographic Guide (Crowood Press, 2011)

Winter, Agnes, and Phythian, Clare

Self-sufficiency Spinning, Dyeing and Weaving (New Holland, 2009) Walsh, Penny

Shepherds and Their Dogs (Merlin Unwin, 2011) Bezzant, John

Showing Sheep (Good Life Press, 2008) Kendrick, Sue

The Complete Book of Tanning Skins and Furs (Stackpole Books, 1987) Churchill, James

The Modern Shepherd (Farming Press, 2002) Brown, Dave, and Meadowcroft, Sam

The Shepherd's Life (Penguin, 2015) Rebanks, James

The Spinner's Guide to Fleece (Storey, 2014) Smith, Beth

The Ultimate Guide to Skinning and Tanning (Lyons, 2002) Burch, Monte

The Yorkshire Shepherdess (Sidgwick & Jackson, 2014) Owen, Amanda

Talking Sheepdogs: Training Your Working Border Collie (Good Life Press, 2008) Scrimgeour, Derek

Weaving Made Easy (Interweave, 2015) Gipson, Liz

DVDs

The Smallholder Series
www. smallholderseries.co.uk
Titles include: Establishing Your Flock; Managing Your Flock for Peak Health; The Breeding Flock; Sheep for Business, Enterprise & Profit; A Guide to Sheep Butchery; A Guide to Showing Sheep.

First Pasture Farm & Country films
www.firstpasture.co.uk
Titles include: Blood, Sweat & Shears; Lamb Survival; Sheep Breeds on the Farm.

Farming and smallholding publications and websites

Accidental Smallholder
www.accidentalsmallholder.net
Country Smallholding
www.countrysmallholding.com
Farmers Guardian
www.farmersguardian.com
Farmers Weekly www.fwi.co.uk
Home Farmer www.homefarmer.co.uk
Meat Trade Journal
www.meatinfo.co.uk
Practical Sheep, Goats & Alpacas www. sgamagazine.co.uk
River Cottage www.rivercottage.net
Sheep Farmer (NSA members' magazine) www.nationalsheep.org.uk
Sheep 101 www.sheep101.info
Sheep Site www.thesheepsite.com
Smallholder www.smallholder.co.uk

Training courses
Husbandry, lambing, shearing
British Wool Marketing Board
Tel: 01274 688666
www.britishwool.org.uk
Farm Skills 01765 608489
www.farmskills.co.uk
Humble by Nature Tel. 01600 714 595
www.humblebynature.com
Kate's Country School www. katescountryschool.co.uk
LANTRA Tel. 02476 696996
www.lantra.co.uk
Smallholder Training Tel. 01837 810569
www.smallholdertraining.co.uk
Sheepdog training/handling
www.sheepdogtraining.co.uk
www.workingsheepdog.co.uk

Breed contacts

Badger Face
Badger Face Welsh Mountain Sheep Society
www.badgerfacesheep.co.uk
Balwen
Balwen Welsh Mountain Sheep Society
www.balwensheepsociety.com
Beulah Speckled Face
Epynt Hill & Beulah Speckled Face Sheep Society www.beulahsheep.co.uk
Berrichon
British Berrichon Sheep Society
www.berrichonsociety.com
Blackface
Blackface Sheep Breeders' Association
www.scottish-blackface.co.uk

Black Welsh Mountain
Black Welsh Mountain Sheep Society
www.blackwelshmountain.org.uk
Bleu du Maine
Bleu du Maine Sheep Society
www.bleudumaine.co.uk
Bluefaced Leicester
Bluefaced Leicester Sheep Breeders' Association www.blueleicester.co.uk
Border Leicester
Society of Border Leicester Sheep Breeders
www.borderleicesters.co.uk
Boreray
Soay Sheep Society
www.soaysheep.org
British Coloured Sheep Breeders' Association
Tel. 01173 771121 www.bcsba.org.uk
British Milksheep
British Milksheep Society
www.britishmilksheep.com
British Rouge
British Rouge Sheep Society
www.rouge-society.co.uk
Cambridge
Cambridge Sheep Society
www.cambridge-sheep.org.uk
Castlemilk Moorit
Castlemilk Moorit Society
www.castlemilkmoorit.co.uk
Charmoise Hill
Charmoise Hill Sheep Society
www.charmoisesheep.co.uk
Charollais
British Charollais Sheep Society
www.charollaissheep.com
Cheviot
Cheviot Sheep Society
www.cheviotsheep.org
North Country Cheviot Sheep Society
www.nc-cheviot.co.uk
Brecknock Hill Cheviot Society
www.brecknockhillcheviotsheep.co.uk
Clun Forest
Clun Forest Sheep Breeders Society
www.clunforestsheep.org.uk
Cotswold
Cotswold Sheep Society
www.cotswoldsheepsociety.co.uk
Dalesbred
Dalesbred Sheep Breeders' Association
www.dalesbredsheep.co.uk
Dartmoor
Grey faced Dartmoor Sheep Breeders' Association
www.greyface-dartmoor.org.uk

WhiteFace Dartmoor Sheep Breeders Association
www.whitefacedartmoorsheep.org.uk

Derbyshire Gritstone
Derbyshire Gritstone Sheep Breeders' Society www.derbyshiregritstone.org.uk

Devon & Cornwall Longwool
Devon & Cornwall Longwool Flock Association www.devonandcornwalllongwool.co.uk

Devon Closewool
Devon Closewool Sheep Breeders www.devonclosewool.co.uk

Dorper
British Dorper Sheep Society www.dorpersheepsociety.co.uk

Dorset Down
Dorset Down Sheep Breeders' Association www.dorsetdownsheep.org.uk

Dorset (Horn/Polled)
Dorset Horn & Poll Dorset Sheep Breeders' Association www.dorsetsheep.org.uk

Easy Care
Easy Care Sheep Society www.easycaresheep.com

Exlana
Sheep Improved Genetics Ltd. www.sig.uk.com

Exmoor Horn
Exmoor Horn Sheep Breeders' Society www.exmoorhornbreeders.co.uk

Gotland
Gotland Sheep Society www.gotlandsheep.com

Hampshire Down
Hampshire Down Sheep Breeders' Association www.hampshiredown.org.uk

Hebridean
Hebridean Sheep Society www.hebrideansheep.org.uk

Herdwick
Herdwick Sheep Breeders' Association www.herdwick-sheep.com

Hill Radnor
Hill Radnor Flock Book Society www.hillradnor.co.uk

Icelandic
British Icelandic Sheep Society www.bisbg.org.uk
Icelandic Sheep in Wales www.icelandicsheepinwales.co.uk

Jacob
Jacob Sheep Society www.jacobsheepsociety.co.uk

Kerry Hill
Kerry Hill Flock Book Society www.yvonnebrown.vpweb.co.uk

Leicestershire Longwool
Leicestershire Longwool Sheep Breeders' Association www.llsba.co.uk

Lincoln Longwool
Lincoln Longwool Sheep Breeders' Association www.lincolnlongwools.co.uk

Llanwenog
Llanwenog Sheep Society www.llanwenog-sheep.co.uk

Lleyn
Lleyn Sheep Society www.lleynsheep.com

Lonk
Lonk Sheep Breeders' Association www.lonk-sheep.org

Manx Loaghtan
Manx Loaghtan Sheep Breeders' Group www.manxloaghtansheep.org

Masham
Masham Sheep Breeders' Association www.masham-sheep.co.uk

Meatlinc
Meatlinc Sheep Company www.meatlinc.co.uk

Norfolk Horn
Norfolk Horn Breeders Group www.norfolkhornsheep.co.uk

North of England Mule
North of England Mule Sheep Association www.nemsa.co.uk

North Ronaldsay
North Ronaldsay Sheep Fellowship www.nrsf.co.uk

Ouessant
Ouessant Sheep Society of Great Britain www.ouessantsheep.net

Oxford Down
Oxford Down Sheep Breeders' Association www.oxforddownsheep.org.uk

Portland
Portland Sheep Breeders' Group www.portlandsheep.co.uk

Romney
Romney Sheep Society www.romneysheep.uk.com

Rough Fell
Rough Fell Sheep Breeders' Association www.roughfellsheep.com

Roussin
Roussin Sheep Society www.bohdgaya.net

Ryeland
Ryeland Flock Book Society www.ryelandfbs.com

Scotch Mule
Scotch Mule Association www.scotchmule.co.uk

Shetland
Shetland Sheep Society www.shetland-sheep.org.uk

Shropshire
Shropshire Sheep Breeders' Association & Flock Book Society www.shropshire-sheep.co.uk

Soay
Soay Sheep Society www.soaysheep.org

Southdown
Southdown Sheep Society www.southdownsheepsociety.co.uk

Suffolk
Suffolk Sheep Society www.suffolksheep.org

Swaledale
Swaledale Sheep Breeders' Association www.swaledale-sheep.com
Teeswater
Teeswater Sheep Breeders' Association www.teeswater-sheep.co.uk

Texel and Beltex
Texel Sheep Society www.texel.co.uk
Beltex www.beltex.co.uk

Valais Blacknose
Valais Blacknose Sheep www.valaisblacknosesheepuk.com

Vendéen
British Vendéen Sheep Society www.vendeen.co.uk

Welsh Halfbred
Welsh Halfbred Sheep Association www.welshhalfbredsheep.co.uk

Welsh Hill Speckled Face
Welsh Hill Speckled Face Sheep Society www.welshhillspeckledface.weebly.com

Welsh Mountain
Welsh Mountain Sheep Society www.welshmountainsheep.co.uk

Welsh Mule
Welsh Mule Sheep Breeders' Association www.welshmules.co.uk

Wensleydale
Wensleydale Longwool Sheep Breeders' Association www.wensleydale-sheep.com

Whitefaced Woodland
Whitefaced Woodland Sheep Society www.whitefacedwoodland.co.uk

Wiltshire Horn
Wiltshire Horn Sheep Society www.wiltshirehorn.org.uk

Zwartbles
Zwartbles Sheep Association www.zwartbles.org

INDEX